Constructive Analysis and Synthesis of Programs

Marco Benini

This document was prepared with LaTeX and reproduced by Lulu.com from a camera-ready PDF copy supplied by the author.

Published by Lulu.com

ISBN 978-1-4452-0638-7

Preface

The idea of using constructive logics to derive computer programs is not new. I learnt about this possible approach during my university years, when I attended a course in Constructive Logics held by Prof. Pierangelo Miglioli. Then, I took my master degree with a dissertation on program verification. Later, during my PhD period, I came again on the use of constructive logics in program verification, analysis and synthesis. At the end, my PhD thesis was about this topic.

When I gained a stable academic position, I went away from the research work made during the PhD. I felt the need for a change and I joined a newly founded Computer Science department where it was required to apply my knowledge to real-world problems. But, I felt that what I did in my PhD dissertation was, in a way, incomplete.

For this reason, when I have been invited in Munich under the Erasmus program, I proposed a graduate course on the use of constructive logics to verify, analyse and synthesise programs. I was aware that a strong group, held by Prof. Helmut Schwichtenberg, is active in the Mathematisches Institut of the Ludwig Maximilian Universitaet in Munich (DE), and I was bold (or, perhaps, crazy) enough to get there and to show a different way to use the same instruments.

This book derives from the lecture notes I wrote as a support for my course. By the way, the course was successful, that is, I began to cooperate with some colleagues in Munich.

Now, I'm continuing to work in the field of Constructive Mathematics, although not anymore in the application

i

on computer programs. And I do that with my colleague and friend, Prof. Peter Schuster, who works in Munich. This book wants to close this long period of my scientific life about the use of constructive systems applied to computer programs.

Most of the material of this book comes from the publications of Pierangelo Miglioli and his group, and from my PhD dissertation. Some of the results shown here have never been published, therefore these notes have been written as a self-contained reference to the described approach.

A warning to the reader is in order: these notes do not try to compare our results to the other ones present in literature. This choice has been done on purpose: I do not want to convey the wrong idea that the illustrated approach is the best one, but I want to illustrate the modus operandi in a pure setting, showing how it naturally originates from a particular point of view about the way to interpret programs.

The reader should understand that the chosen point of view is by no means privileged, in fact, other solutions are possible and fruitful. But the reader will find that the described style to the debated problems is of interest since it provides an explanation to the formal analysis of programs.

The prerequisites to the understanding of the content of this work are: a general knowledge of mathematical logic and its techniques and some basic skills in a programming language. Moreover, an experience in Formal Methods and in the application of constructive mathematics is helpful, although not strictly required.

Contents

Chapter 1

Introduction

This chapter wants to introduce the problems of verification, analysis and synthesis of computer programs in a formal environment based on constructive mathematics. In particular, we want to bring the reader to consider these problems from a point of view where it becomes clear that a constructive reasoning system is the best model of the way of thinking of a programmer. Thus, it becomes natural to prove the correctness of a program (verification task) in a way which follows the reasoning the programmer used to code the program.

Moreover, we will discuss the value of a correctness proof from a social/economical point of view: we mean that a correctness proof cannot reasonably be limited to an object saying whether the program behaves as expected or it contains some sort of mistakes. A correctness proof should be used to convince an expert in programming that the program works as expected, and it should provide an understandable explanation of the reason. In addition, the proof should provide a series of collateral information which allows the programmer to extend his knowledge of the program behaviour, permitting to reuse (parts of) the program in different contexts without proving again its properties in the new setting.

Finally, the specification, which encodes what a program is expected to compute, should be rich enough to allow a computational reading beyond the purely logical meaning. In fact, a specification can be either a statement (the program has to satisfy this property) or a procedure (the program has to produce its results in this way).

1.1 How programmers think to their code

A programmer writes programs, functions and algorithms according to an informal specification, i.e., to a description of the problem to solve along with a series of constraints on the admissible solutions. Precisely, the programmer design algorithms that, in his opinion, solve the problem and, then, he codes the algorithms in a suitable programming language.

Since algorithms as well as their coding may contain errors, that is, they do not obey to the specification, it is required some evidence of the correctness of algorithms and of their implementation.

The most popular way to provide an evidence of the correctness of a program is *testing*: practically, we show that in a number of significant cases the program behaves as expected, thus we infer that the program is correct. Obviously, testing has an important value, but it does not guarantee that the program is correct; in fact, it proves beyond any doubt that the program is correct in the test cases, and it justifies the hope that it behaves correctly in the similar cases. If the test coverage is wide enough, there is a good confidence in the correctness of the program in every case.

Sometimes, when the programs operate in a critical environment, e.g., air traffic control, or when a programming error has important economical consequences, e.g., microprocessors' firmware, testing is not sufficient. The discriminant between applications where testing can be employed and applications where testing is not enough lies in the idea of *acceptable error*: whenever the consequences of an error are not tolerable, then testing is not enough.

In fact, testing cannot assure the absence of errors, but it can just enhance the confidence that, if there is an error, it will manifest in a rare case. The reason lies in the fact that only *exhaustive testing* can find every error, and few programs have a finite set of possible inputs, and even less have a small (tractable) set of possible inputs, thus exhaustive testing is, in general, unfeasible.

When testing is not enough, we need another kind of program verification: *formal verification*, i.e., we want to prove in a formal, mathematical way, the *correctness proof*, that the program satisfies the specification.

There are a number of practical problems in formal verification: in order to reason about the program behaviour, we need to transform the program itself into a mathematical object, its *representation*; to prove that the mathematical representation of the program satisfies the specification we need to formalise the specification as a mathematical object, and we have to explain in mathematical terms what we intend with the verb *to satisfy*.

Usually, the problem of program formalisation, is solved by restricting the programming language in such a way that the admissible programs have a well-understood formal semantics. On the contrary, the second problem, formalising specifications, is rarely considered, assuming that, when formal verification is required, the problem is described from the very beginning in mathematical terms. The notion of satisfiability is decided a priori on a logical base: the program representation satisfies the specification if we can prove the formal specification assuming the program representation in some fixed logical system.

As a matter of fact, most correctness proofs are difficult to read because of their length and their inelegance, but their structure is somewhat similar to the informal reasoning the programmer uses to convince himself about the correct behaviour of his own code. Thus, the programmers' arguments are not *proofs*, but they are used as heuristics to guide the formal proving process.

Henceforth, we believe that trying to understand the point of view of programmers is important in the formal verification task. Observing the way of thinking of programmers we can learn how to perform better proofs of their programs, because we understand how specifications are implemented and what the programmer believes to hold in the critical points of his code. Our observation has shown that programmers think in a double way:

- when reasoning about data, a programmer takes a classical point of view: a statement about data is either true

function quicksort(a: array) { sort(a, 1, length(a)); }

function sort(a: array, l, r: int) {
    $x := a[(l + r)$ div $2]$; $i := l$; $j := r$;
    repeat
        while $(a[i] < x)$ $\{i := i + 1; \}$;
        while $(a[j] > x)$ $\{j := j - 1; \}$;
        if $(i < j)$ **then** $\{w := a[i]; a[i] := a[j]; a[j] := w; \}$
        $i := i + 1; j := j - 1$;
    until $(i \geq j)$;
    if $(l < j)$ **then** sort(a, l, j);
    if $(i < r)$ **then** sort(a, i, r); }

Figure 1.1: The quicksort procedure.

or false. For example, a specific element occurs or it does not in a database; a list is ordered or it is not; a pointer contains a valid reference or it does not.

- when reasoning about code, a programmer takes a constructive point of view: the program is written as a computable *construction* of the solution. This amount to say, for example, that it does not suffice to show that there is a particular element in a database, but the program must encode a way to find it.

This observation is better understood by means of a concrete example. Let us look at Figure 1.1: the depicted function sorts an array of elements, and the underlying algorithm is *quicksort* as found in many textbooks, see. e.g., [10].

As programmers, we think as follows: we choose an element x in the array, and we exchange elements in the array so that every element less than x appears before any element greater than x. The final picture after the **repeat** ... **until** loop (its *postcondition*) is that the array is partitioned into two subarrays: the first one contains all the elements less than x, the other one contains all the element greater than x. Then, we sort these two subarrays and, thus, the final array gets sorted.

We want to remark two important points:

- we explicitly construct a bi-partition of the array a in the sort function by a computational procedure represented in the **repeat** ... **until** loop;

- in every test, that is, when evaluating the conditions in the **while**, **if** and **until** statements, we assume that if the condition does not hold then its negation is true.

In other words, a correctness proof showing that the array a becomes sorted after the application of the quicksort function, must prove that the partitioning loop constructs, indeed, a partition as explained. During the development of the correctness proof we are allowed to assume that, whenever a test fails, the negation of its condition holds.

Thus, we are using classical logic when we reason on data, since we assume that a condition must be either true or false, and, if it is not true, then it must be false, a clear instance of the *excluded middle principle*, also known as *tertium non datur*.

An immediate consequence of the correctness proof is the knowledge that every array can be sorted. In a sense, this statement is the classical content of our correctness proof: we can easily prove the same result by showing that there is a permutation of the elements of the array that produces a sorted array by a clever use of the pigeon hole principle.

However, a proof that shows that every array can be sorted has no use, unless it shows also how to construct the sorted array. Namely, the correctness proof of quicksort underlies a construction, conventionally identified with the quicksort algorithm, that shows how to build the sorted array.

From a purely mathematical point of view, the correctness proof for quicksort, intended as a proof of the existence of a sorted array, is unnecessarily complex and it also has a bizarre and inelegant structure; on the contrary, from the verification point of view, its complexity is required to show that a program is correct, or, from a different perspective, the proof is itself an interesting computable construction that allows to establish the truth of the existential statement.

When a programmer writes his code, he justifies its construction by means of an informal correctness argument that respects the foundation of constructive reasoning: every truth he establishes, must have a justification that is a construction. Whenever an existential postcondition holds in his program, the programmer claims its truth because the preceding code calculates a *witness* for the "there exists" statement.

For example, after the main loop in the quicksort function, we can claim that there exists a bi-partition of the a array such that every element less than x lies in the first equivalence class, and every element grater than x lies in the second equivalence class. This statement is essential to prove the correctness of the sort function since we have to prove that the whole array gets sorted, assuming that the two recursive function calls at the end really sort the two partitions.

However, an acceptable proof of the existence of the bi-partition must show how to compute it, otherwise we cannot check the code that forms our quicksort implementation.

1.2 The economical value of verification

As a matter of fact, history of formal verification [26] has shown that proving the correctness of real computer programs is very difficult and very expensive: in fact, a large number of experts is required to work for a long time.

The immediate consequence of this cost has been the limitation of formal verification to absolutely critical tasks, e.g., nuclear plants, critical aspects of space exploration, strategic military applications, ...

In our opinion, there is another problem rising from the cost of formal verification: as a general rule of business, something is convenient if the benefits it produces are more important than the price to pay. In the case of classical formal verification, the correctness proof is expensive, but its value is limited: indeed, its immediate application produces just the warranty that the verified program is correct in every case, on the whole input space. In a simplistic way, the correctness proof is just a boolean answer to the question "Is the program correct?". The whole process is a very expensive effort to obtain for just one bit of information, isn't?

Thus, since the correctness proof is a mathematical justification of the truth of the statement "the program behaves as expected", it should be possible to use the correctness proof to establish other interesting facts about the verified program. In other words, the correctness proof exposes many true facts about the verified program, and the knowledge of these facts can motivate someone to invest in the proof development.

Obviously, we may expect someone to invest in an expensive process if, and only if, there is "contract" that guarantees that the set of facts about the program that we will produce is "significant" both in the number and in the insight they give to a better understanding on the program behaviour.

Hence, a set of facts is interesting if it allows us to better understand how the program works, and so to reuse some parts of the program in a different application, being sure that the new employment does not require to prove again the correctness of those parts in the new setting.

In addition, a better understanding of the program is very useful to allow an human expert on the problem domain to "certify" that the formal specification really corresponds to what the program is intended to do. In fact, in the real world, few specifications are formal in the very beginning, even in critical applications, and thus it is important to have some sort of feedback from the correctness proof that what has been proved is, indeed, what was intended to be.

1.3 Toward a reasonable formalisation

The discussion in the first two sections of this chapter clarifies a number of requirements on the formal verification process: it should permit to follow the way to reason of programmers, it should "think" to data in a classical way, and to programs in a constructive way[1], and it should build correctness proofs that can be analysed and manipulated to extract useful information and knowledge about the verified programs.

Moreover, the language used to formalise specifications should be able to code two ways of specifying requirements: a logical one, i.e., "this fact must hold!", and a procedural one, i.e., "it must compute this!". The specification language, and thus the reasoning system must smoothly combine these two aspects when the correctness proof is developed.

In this monograph we will see that it is possible to satisfy all these requirements when choosing an appropriate rea-

---

[1]The reader is invited notice that our use of the word "constructive" is, at the moment, very distant from the usual meaning it gets in Mathematics.

soning system. We should warn the reader that our solution has a limited practical application due to the computational complexity of our techniques and methods. Despite this limitation, from a purely theoretical point of view, the approach we will describe is well founded, allowing, at least in principle, to formally verify the correctness of computer programs in a "deeper" way, that enables the extraction of useful information from correctness proofs, to specify the program behaviour both in a logical and in a computational way, to model data in classical logic and, at the same time, to reason about programs with a strictly constructive attitude, and much more.

The starting point of our approach is to consider a reasoning system which is constructive in the mathematical sense, and to interpret its constructive character as a computational reading for its formulae, and thus, for specifications. Almost automatically, we get an interpretation of proofs as programs.

In addition, a clever formalisation of data types allows to reason on them in a constructive logical system as if they were classically conceived, because their semantics is still classical when embedded into the constructive reasoning system.

Finally, by devising an appropriate definition of constructive system, slightly out of the established tradition, we gain the possibility to extract information from formal proofs. Furthermore, we will show some properties about the amount and the quality of the information contained in a proof which can be effectively extracted by our procedures. In fact, the illustration of these results will be our core result.

1.4 A sensible reasoning system

The basic requirement we have on a reasoning system is about its expressiveness: we want to write specifications that can be read both as statements and as computational constraints.

Precisely, if a specification is a formula in a logical language, we want that, beside the usual meaning induced by semantics, a formula has also a computational meaning. Practically, we require that

- the formula $A \wedge B$ is true iff A is true and B is true (logical reading), and $A \wedge B$ represents the concurrent execution of A and B (computational reading);

- the formula $A \vee B$ is true iff either A is true or B is true (logical reading), and $A \vee B$ represents the computation that chooses[2] and executes one of A and B (computational reading);

- the formula $\exists x. A(x)$ is true if there exists an element e in the universe, such that, interpreting x as e, the formula $A(x)$ is true (logical reading), and $\exists x. A(x)$ represents a computation whose inputs are the free variables in $A(x)$ except for x, and whose output is an element e that makes $A(x)$ logically true, and which is computed according to the computational meaning of $A(x)$.

We silently assume that the atomic formulae represent basic computations, like the expressions in a programming language. In addition, we require the reasoning system assigns a computational meaning to every connective and quantifier in the logical language although most specifications do not make use of the full language, as we will see in the following.

From a mathematical perspective, we may ask whether it is possible to construct a logical system admitting also a computational reading for formulae. As we will see, the answer is positive. Nevertheless, we prefer to start the discussion of our reasoning system by showing why a system based on classical logic is not admissible. In fact, a reasoning system where both a logical and a computational reading are admissible must provide a semantics that allows to interpret formulae in both ways, and classical logic (**CL** from now on) does not.

Specifically, let us analyse the specification $A(t) \vee \neg A(t)$ where $A(x)$ is the formula coding the property "the program x terminates on every input". It is a well-known fact that this property can be coded in the standard Peano arithmetic, modulo a Gödelisation of programs, as any good textbook on Recursion Theory shows, see, e.g., [45].

But the formula $A(t) \vee \neg A(t)$ is trivially true in **CL**, in fact, its proof follows immediately as a consequence of the excluded middle principle: for every formula B, it holds that $B \vee \neg B$. Thus, the computational reading of $A(t) \vee \neg A(t)$ is

---

[2]The choice action is important: we require that it is always possible to know what choice has been done, as it happens in any real program.

that either we can calculate that the program t terminates on every input, or we can decide that there is at least an input such that t does not terminate. Although the logical reading is obviously true in classical logic, the computational reading cannot be decided in general, because the decision problem whose instances are the computable functions and whose question is "is the instance a total function?" is an example of non-computable problem.

So, the computational meaning of $A(t) \vee \neg A(t)$ in **CL** is not an effective computation and it cannot be coded by any program. Moreover, even if we accept this odd notion of computational reading, we cannot find in the proof of the formula any information whether t computes a total function or not, since the proof is completely non-informative on this aspect, being just an instance of an axiom, the one representing the excluded middle principle.

On the contrary, if we drop the excluded middle principle and we try to prove $A(t) \vee \neg A(t)$, we are forced to show whether the program t terminates on every input or not. In other words, we must perform a specific proof for the program t and, hopefully, this proof will contain enough information to decide whether t computes a total function or not.

Abstracting from the specific example, classical logic, **CL**, is incompatible with a computational reading of its formulae because it does not construct a recursive decision procedure for its disjunctive theorems.

Henceforth, a sensible reasoning system for or purposes must be non-classical, and it must construct the witnesses for its disjunctive and existential theorems, that is,

- whenever it proves a theorem of the form $A \vee \neg A$, it has to provide enough information to know whether A holds or $\neg A$ holds, that is, it must be that either A is a theorem or $\neg A$ is;

- whenever it proves a theorem of the form $\exists x. A(x)$, it has to provide enough information to show a term t such that $A(t)$ holds, that is, $A(t)$ must be a theorem.

We will speak of a *naïvely constructive logical system* when a logical system or theory T is such that

- **[disjunction property]** if T proves $A \vee \neg A$, then either T proves A or T proves $\neg A$;

- **[explicit definability property]** if T proves the theorem $\exists x. A(x)$, then there is a term t such that T proves $A(t)$.

It is a fact, see, e.g., [51], that the notion of naïvely constructive system does not capture the essence of constructive mathematics, and many recognised constructive systems violate the rather strict requirement of satisfying the disjunction property and the explicit definability property.

Nevertheless, the introduced notion seems, at a first sight, to capture in a simple and natural way the systems allowing a computational reading of formulae, as far as the logical system is decidable. We will return in Chapter 7 on this point, showing that the notion of naïvely constructive logical system is insufficient for our purposes since it does not allow to model in a strong sense the computational reading of formulae and proofs. Specifically, we want some bounds on the complexity to discover the witnesses with respect to the complexity to prove the formula: naïvely constructive systems fail to meet any bound.

As the reader have noticed, we gave a sketchy presentation of the computational reading of formulae, enough to discuss why classical systems are inadequate for our purposes. Although the definition of the computational meaning can be extended to include implication and universal quantification, these cases are unproblematic since the intended interpretation of connectives in a constructive system, the so-called BHK interpretation [49], which is often cited as the "intended" meaning of intuitionistic logic, forces an essentially unique computational reading.

A problem arises when we consider negation: in fact, the intuitionistic negation is definable in the system as $\neg A \equiv A \rightarrow \bot$, thus the negation of a fact A holds when the fact leads to a contradiction, $\bot$. Therefore, the computational interpretation of $\neg A$ becomes a statement saying that A is impossible be calculate. Alternatively, and more positively, one may consider a constructive interpretation of negation: $\neg A$ holds if there exists a counterexample to the truth of A, and the counterexample is a construction of the truth of $\neg A$. Actually, it is possible to formalise a constructive notion of negation, following, e.g., [44].

In the rest of this book we will focus on two logical systems: intuitionistic logic, **IL** from now on, and the **E** logic. The latter adopts a constructive negation and, thus, has a computational reading of formulae where negation is interpreted as a search for a counterexample. Both these logical systems are naïvely constructive, of course.

The last requirement we have discussed is the ability to reason in a classical way on data and in a constructive way on programs. Since we have chosen to work in two naïvely constructive logical systems, half of the requirement, the constructive part, is satisfied.

As we will discuss in Chapter 3, it is possible to extend a logical system with a theory that models the required data types and, at the same time, that preserves the constructive character of the logic. Moreover, the theory has a clear classical meaning that gets preserved when interpreted in a suitable constructive system. The reader will not wonder when discovering that **IL** and **E** are "suitable". Moreover, it will not be astonishing to discover that "suitable" in this context means uniformly constructive, our strengthening of the concept of naïvely constructive.

In addition, since the formulae in our logical systems have both a computational and a logical reading, a specification can be written considering both meanings, and thus, it is possible to specify both the required actions (computational reading) and the properties to hold (logical reading).

We can and we do require something subtler: we want to be able to specify that a fact holds "when it is interpreted in classical logic". This requirement make sense in our problem, since we think to data as if the logical system would be classical, and thus it becomes natural to declare that a specification on data has no computational meaning, but it has to be interpreted in a strictly classical sense. But, the reader has to notice that we do not limit our requirement to formulae dealing with data.

For this very reason we use the **E** logical system, where a special unary operator on formulae is present: $\Box$, spelt "Box". The meaning of $\Box A$ is: the formula $\Box A$ is true if A is true in

CL. The computational meaning of $\Box A$ is the elementary computation terminating on every input with no output. So, $\Box$ acts as a logical wrapper for classically interpreted formulae, while it is transparent to the computational reading, acting as a no-operation requirement.

Chapter 2

Logical Systems

In this chapter, we will introduce the first-order intuitionistic logic and the E logic from a mathematical point of view.

In Chapter 1 we expressed our interest for constructive systems where the disjunction property and the explicit definability property hold. This interest raised since we have shown that these properties do not hold in classical logic and they enable us to read formulae and proofs as objects with a computational meaning.

The purpose of this chapter is double: in the first place, we want to introduce the logical systems we will use to explore how far the intuition of Chapter 1 can be brought; in the second place, we want to state and to prove the fundamental properties that will be used in our study.

Most of the results about intuitionistic logic will be stated without proofs, since a rich literature is available; in these cases, the bibliographic references are used to point to the best treatment in literature, according to the aim the result has been introduced for, and to the (debatable) author's opinion.

2.1 The intuitionistic logic

The intuitionistic first-order logic (**IL** for short) is the most well-known constructive system. Among the many references that are available about this system, we prefer to signal [47, 48] to whom is interested in a wider presentation.

In this work, we will use the natural deduction presen-

$$\frac{A \quad B}{A \wedge B} \wedge I \qquad \frac{A \wedge B}{A} \wedge E_l \qquad \frac{A \wedge B}{B} \wedge E_r$$

$$\frac{A}{A \vee B} \vee I_l \qquad \frac{B}{A \vee B} \vee I_r \qquad \frac{A \vee B \quad \overset{[A]}{\underset{\vdots}{C}} \quad \overset{[B]}{\underset{\vdots}{C}}}{C} \vee E$$

$$\frac{\overset{[A]}{\underset{\vdots}{B}}}{A \rightarrow B} \rightarrow I \qquad \frac{A \quad A \rightarrow B}{B} \rightarrow E$$

$$\frac{\overset{[A]}{\underset{\vdots}{C}} \quad \overset{[A]}{\underset{\vdots}{\neg C}}}{\neg A} \neg I \qquad \frac{\neg A \quad A}{B} \neg E \; (**)$$

$$\frac{A(p)}{\forall x. A(x)} \forall I \; (*) \qquad \frac{\forall x. A(x)}{A(t)} \forall E$$

$$\frac{A(t)}{\exists x. A(x)} \exists I \qquad \frac{\exists x. A(x) \quad \overset{[A(p)]}{\underset{\vdots}{B}}}{B} \exists E \; (*)$$

where, in $(*)$, p is an eigenvariable, and in $(**)$, B is an atomic formula.

Table 2.1: Inference rules for **IL**.

tation of **IL**, as shown in Table 2.1: this calculus is almost the standard one, as found in literature. Two main subtleties should be remarked:

- Usually, **IL** is based on a language that contains the $\perp$ (false) constant as primitive, while the $\neg$ (not) operator is defined as $\neg A \equiv A \to \perp$. We prefer the calculus where $\neg$ is a primitive operator to easier the comparison with the **E** logic where the $\neg$ operator is primitive and not definable. Moreover, the $\neg$ operator has a computational meaning, thus, having it as primitive, it becomes simpler to discuss how the logical system influences the computational interpretation of formulae and proofs.

- Usually, the $\neg$E inference rule is unconstrained, i.e., its conclusion is allowed to be any formula, not just atomic formulae as in our case. The reason for the constraint is purely technical: it simplifies many proofs and definitions we will introduce in the rest of this work. By induction on the structure of formulae, one may easily prove that the restricted rule is as powerful as the unconstrained one.

Also, the reader should notice that a sound and complete calculus for classical logic (**CL**) is obtained by adding a single rule to **IL**:

$$\frac{\begin{array}{cc} [\neg A] & [\neg A] \\ \vdots & \vdots \\ B & \neg B \end{array}}{A} \neg E_{CL}$$

which is equivalent to the more popular axiom $\neg\neg A \to A$.

Kripke semantics

To aid the comparison between the **IL** and the **E** logic, we define the Kripke semantics for **IL**. One can prove that the **IL** calculus is sound and complete with respect to this semantics[1]. The proof is standard and can be easily found in the already cited literature.

---

[1]Thus, we are justified when speaking about **IL** instead of **IL** logic, **IL** calculus, **IL** semantics, . . .

Specifically, a Kripke model for **IL** is a quadruple $\mathfrak{K} = \langle P, \leq, \iota, D \rangle$, where

- $\mathbb{P} = \langle P, \leq \rangle$ is a partial ordered set, the *frame*;

- D is the *domain* function, associating, to any $\alpha \in P$, a domain $D(\alpha)$ such that, for any $\alpha, \beta \in P$, if $\alpha \leq \beta$ then $D(\alpha) \subseteq D(\beta)$;

- the *evaluation* function ι associates with every $\alpha \in P$ a map from the set of atomic formulae to the set $\{\top, \uparrow\}$ of truth values, where $\uparrow$ stands for undefined[2].

The ι is such that, for every $\alpha, \beta \in P$ and for every atomic formula ϕ, if $\alpha \leq \beta$ and ϕ belongs to the domain of ι_α then ϕ belongs to the domain of ι_β, and $\iota_\alpha(\phi) = \iota_\beta(\phi)$.

The ι function can be extended to arbitrary formulae:

- $\iota_\alpha(\neg A) = \top$ iff, for every $\beta \in P$, $\alpha \leq \beta$, $\iota_\beta(A) = \uparrow$;

- $\iota_\alpha(A \wedge B) = \top$ iff $\iota_\alpha(A) = \top$ and $\iota_\alpha(B) = \top$;

- $\iota_\alpha(A \vee B) = \top$ iff $\iota_\alpha(A) = \top$ or $\iota_\alpha(B) = \top$;

- $\iota_\alpha(A \rightarrow B) = \top$ iff, for every $\beta \in P$, $\alpha \leq \beta$, either $\iota_\beta(A) = \uparrow$ or $\iota_\beta(B) = \top$;

- $\iota_\alpha(\forall x. A(x)) = \top$ iff, for every $\beta \in P$, $\alpha \leq \beta$, for every $c \in D(\beta)$, $\iota_\beta(A(c)) = \top$;

- $\iota_\alpha(\exists x. A(x)) = \top$ iff there exists a $c \in D(\alpha)$ such that $\iota_\alpha(A(c)) = \top$.

One can easily verify that, for every formula A, if $\alpha \leq \beta$, and $\iota_\alpha(A) \neq \uparrow$, then $\iota_\alpha(A) = \iota_\beta(A)$.

The notion of truth in **IL** is captured by:

Definition 2.1.1 *A formula A is* valid *in the model $\mathfrak{K}$, notation $\vDash_\mathfrak{K} A$, iff, for every $\alpha \in P$, $\iota_\alpha(A) = \top$; a formula is* valid *(in **IL**), notation $\vDash A$, iff it is valid in every model.*

A formula A is a consequence *of a set of formulae Γ in the model $\mathfrak{K}$, notation $\Gamma \vDash_\mathfrak{K} A$, iff for every $\gamma \in \Gamma$, such that $\vDash_\mathfrak{K} \gamma$, it holds $\vDash_\mathfrak{K} A$; a formula A is a* consequence *of Γ (in **IL**) iff, for every model $\mathfrak{K}$, $\Gamma \vDash_\mathfrak{K} A$.*

---

[2]The ι function is equivalent to the more usual forcing relation.

The Soundness Theorem states that, whenever A is deduced by Γ in the **IL** calculus, then it is a consequence of Γ:

Theorem 2.1.1 (Soundness) *If* $\Gamma \vdash A$ *then* $\Gamma \vDash A$.

The Completeness Theorem states that, whenever A is a consequence of Γ, it is possible to prove A from Γ:

Theorem 2.1.2 (Completeness) *If* $\Gamma \vDash A$ *then* $\Gamma \vdash A$.

The proofs of the Soundness Theorem and of the Completeness Theorem can be found in [50]. We want to remark that the proof of the Completeness Theorem is not constructive: the proof shows that, if $\Gamma \nvdash A$, then there is a Kripke model $\mathfrak{K}$ such that $\vDash_{\mathfrak{K}} \Gamma$ and $\nvDash_{\mathfrak{K}} A$.

Proof-theoretical results

The proof theory of **IL** is vast and complex. In the author's opinion, the most approachable text is [47].

The very first result we want to show is that **CL**, classical logic can be embedded into the **IL** system.

Theorem 2.1.3 *There exists a map* τ *from classical formulae to intuitionistic formulae such that* $\{\tau(\gamma) \mid \gamma \in \Gamma\} \vdash_{\text{IL}} \tau(A)$ *iff* $\Gamma \vdash_{\text{CL}} A$.

Proof: The proof can be found in , e.g., [47]. One of the possible definition for τ is the negative translation: $\tau(p) = \neg\neg p$, for every atomic formula p, $\tau(\neg(A)) = \neg\tau(A)$, $\tau(A \wedge B) = \tau(A) \wedge \tau(B)$, $\tau(A \vee B) = \neg(\neg\tau(A) \wedge \neg\tau(B))$, $\tau(A \to B) = \tau(A) \to \tau(B)$, $\tau(\forall x. A(x)) = \forall x. \tau(A(x))$ and $\tau(\exists x. A(x)) = \neg\forall x. \neg\tau(A(x))$. $\qquad\square$

Thus, in a limited sense, what is classically valid, can be proved in **IL**, modulo a translation. Obviously, the translation is transparent from a classical point of view, that is, $\tau(A) \leftrightarrow A$ in **CL**, but it is evident that the constructive content of the classical statement is completely destroyed in translation process.

The second result we want to introduce is the subformula property for **IL**. It states that, for every proof in normal form of $\Gamma \vdash_{\text{IL}} A$, every formula in the proof is a subformula of a formula in $\Gamma \cup \{A\}$. We will not prove the statement, nor we will clarify what is intended by *proof in normal form*, being these concepts beyond the scope of this notes. The reader is invited to see [47] for a discussion.

From our point of view, the interesting fact we can deduce from the subformula property is that, by construction, a proof in normal form of a non atomic formula ends with the application of an introduction inference rule.

Thus, let us suppose that $\vdash_{IL} \exists a. A(x)$, by means of the proof π in normal form. Then, the last step in the proof is an introduction rule, thus, for some term t,

$$\pi \equiv \begin{array}{c} \vdots \\ \exists x. A(x) \end{array} \equiv \cfrac{\begin{array}{c} \vdots \\ A(t) \end{array}}{\exists x. A(x)} \exists I \; .$$

Hence, we have a proof of $\vdash_{IL} A(t)$, for some term t, showing that **IL** has the explicit definability property.

Similarly, let us suppose that $\vdash_{IL} A \vee B$, by means of the proof π in normal form. Then, the last step in the proof is an introduction rule, thus

$$\pi \equiv \begin{array}{c} \vdots \\ A \vee B \end{array} \equiv \cfrac{\begin{array}{c} \vdots \\ A \end{array}}{A \vee B} \vee I \quad \text{or,} \quad \pi \equiv \begin{array}{c} \vdots \\ A \vee B \end{array} \equiv \cfrac{\begin{array}{c} \vdots \\ B \end{array}}{A \vee B} \vee I \; .$$

Hence, we have either a proof of $\vdash_{IL} A$ or a proof of $\vdash_{IL} B$, showing that **IL** has the disjunction property.

Thus, we conclude that **IL** is naïvely constructive, and, moreover, a witness for disjunctive and existentially quantified formulae can be found in their proofs: it suffices to transform the proof in a normal form proof, and then to look at the last proof step.

2.2 The E logic

The **E** logic has been introduced in [38] as a way to incorporate classical truth within a constructive system. In fact, the **E** system is characterised by the $\square$ operator whose operand, a formula, is classically interpreted, while non-boxed formulae are read in a constructive way.

Moreover, the **E** system has a non-intuitionistic notion of negation, since, as said in Chapter 1, we prefer a constructive notion of negation, a la [44], i.e., whenever we prove a negated formula, it is possible to prove a counterexample.

Tables 2.2 and 2.3 show the natural deduction presentation of the **E** system. As for the **IL** system, the ¬E rule is restricted to atomic formulae in order to simplify proofs; as before, this is a technical trick that does not influence the deductive power of the calculus.

Kripke semantics

A Kripke model for **E** is a quadruple $\mathfrak{K} = \langle P, \leq, \iota, D \rangle$, where

- $\mathbb{P} = \langle P, \leq \rangle$ is a partial ordered set, the *frame*, with the constraint that, for every $\alpha \in P$, there is a $\beta \in P$, with $\alpha \leq \beta$ such that β is *final*, that is, for every $\gamma \in P$, with $\beta \leq \gamma, \beta = \gamma$;

- D is the *domain function* associating, to any $\alpha \in P$, a domain $D(\alpha)$ such that, for any $\alpha, \beta \in P$, if $\alpha \leq \beta$ then $D(\alpha) \subseteq D(\beta)$;

- the *evaluation* function ι associates with every $\alpha \in P$ a map from the set of atomic formulae to the set $\{\top, \bot, \uparrow\}$ of truth values, where $\uparrow$ stands for undefined.

The ι function must satisfy the following conditions:

- for every $\alpha, \beta \in P$ and for every atomic formula ϕ, if $\alpha \leq \beta$ and ϕ belongs to the domain of ι_α then ϕ belongs to the domain of ι_β, and $\iota_\alpha(\phi) = \iota_\beta(\phi)$.

- for every $\alpha \subset P$ and for every atomic formula ϕ, there is a $\beta \in P$ such that $\alpha \leq \beta$ and $\iota_\beta(\phi) \neq \uparrow$.

The ι function can be extended to arbitrary formulae:

- $\iota_\alpha(\neg A) = \top$ iff $\iota_\alpha(A) = \bot$; $\iota_\alpha(\neg A) = \bot$ iff $\iota_\alpha(A) = \top$.

- $\iota_\alpha(A \wedge B) = \top$ iff $\iota_\alpha(A) = \top$ and $\iota_\alpha(B) = \top$; $\iota_\alpha(A \wedge B) = \bot$ iff $\iota_\alpha(A) = \bot$ or $\iota_\alpha(B) = \bot$.

- $\iota_\alpha(A \vee B) = \top$ iff $\iota_\alpha(A) = \top$ or $\iota_\alpha(B) = \top$; $\iota_\alpha(A \vee B) = \bot$ iff $\iota_\alpha(A) = \bot$ and $\iota_\alpha(B) = \bot$.

$$\frac{A \quad B}{A \wedge B} \wedge I \qquad\qquad \frac{\neg A}{\neg(A \wedge B)} \neg \wedge I \qquad \frac{\neg B}{\neg(A \wedge B)} \neg \wedge I$$

$$\frac{A \wedge B}{A} \wedge E \quad \frac{A \wedge B}{B} \wedge E \qquad \frac{\neg(A \wedge B) \quad \overset{[\neg A]}{\overset{\vdots}{C}} \quad \overset{[\neg B]}{\overset{\vdots}{C}}}{C} \neg \wedge E$$

$$\frac{A}{A \vee B} \vee I \quad \frac{B}{A \vee B} \vee I \qquad \frac{\neg A \quad \neg B}{\neg(A \vee B)} \neg \vee I$$

$$\frac{A \vee B \quad \overset{[A]}{\overset{\vdots}{C}} \quad \overset{[B]}{\overset{\vdots}{C}}}{C} \vee E \qquad \frac{\neg(A \vee B)}{\neg A} \neg \vee E \quad \frac{\neg(A \vee B)}{\neg B} \neg \vee E$$

$$\frac{\overset{[A]}{\overset{\vdots}{B}}}{A \rightarrow B} \rightarrow I \qquad\qquad \frac{A \quad \neg B}{\neg(A \rightarrow B)} \neg \rightarrow I$$

$$\frac{A \quad A \rightarrow B}{B} \rightarrow E \quad \frac{\neg(A \rightarrow B)}{A} \neg \rightarrow E \quad \frac{\neg(A \rightarrow B)}{\neg B} \neg \rightarrow E$$

$$\frac{A \quad \neg A}{B} \neg E(*) \quad \frac{A \quad \neg A}{\neg B} \neg E(*) \quad \frac{A}{\neg\neg A} \neg\neg I \quad \frac{\neg\neg A}{A} \neg\neg E$$

$$\frac{\overset{[\neg A]}{\overset{\vdots}{B}} \quad \overset{[\neg A]}{\overset{\vdots}{\neg B}}}{\Box A} \Box I \qquad \frac{\overset{[A]}{\overset{\vdots}{B}} \quad \overset{[A]}{\overset{\vdots}{\neg B}}}{\neg\Box A} \neg\Box I$$

where, in $(*)$, B is any atomic formula.

Table 2.2: Inference rules for the **E** logic (I).

$$\frac{A(p)}{\forall\, x.\, A(x)}\; \forall\mathrm{I}(*) \qquad\qquad \frac{\neg A(t)}{\neg\forall\, x.\, A(x)}\; \neg\forall\mathrm{I}$$

$$\frac{\forall\, x.\, A(x)}{A(t)}\; \forall\mathrm{E} \qquad \frac{\neg\forall\, x.\, A(x) \quad \begin{matrix}[\neg A(p)]\\ \vdots\\ B\end{matrix}}{B}\; \neg\forall\mathrm{E}(*)$$

$$\frac{A(t)}{\exists\, x.\, A(x)}\; \exists\mathrm{I} \qquad\qquad \frac{\neg A(p)}{\neg\exists\, x.\, A(x)}\; \neg\exists\mathrm{I}(*)$$

$$\frac{\exists\, x.\, A(x) \quad \begin{matrix}[A(p)]\\ \vdots\\ B\end{matrix}}{B}\; \exists\mathrm{E}(*) \qquad \frac{\neg\exists\, x.\, A(x)}{\neg A(t)}\; \neg\exists\mathrm{E}$$

where, in $(*)$, p is an eigenvariable.

Table 2.3: Inference rules for the **E** logic (II).

- $\iota_\alpha(A \to B) = \top$ iff, for every $\beta \in P$, $\alpha \leq \beta$, $\iota_\beta(A) = \uparrow$ or $\iota_\beta(A) = \bot$ or $\iota_\beta(B) = \top$; $\iota_\alpha(A \to B) = \bot$ iff $\iota_\alpha(A) = \top$ and $\iota_\alpha(B) = \bot$.

- $\iota_\alpha(\Box A) = \top$ iff, for every $\beta \in P$, $\alpha \leq \beta$, β final, $\iota_\beta(A) = \top$; $\iota_\alpha(\Box A) = \bot$ iff, for every $\beta \in P$, $\alpha \leq \beta$, either $\iota_\beta(A) = \uparrow$ or $\iota_\beta(A) = \bot$.

- $\iota_\alpha(\forall\, x.\, A(x)) = \top$ iff, for every $\beta \in P$, $\alpha \leq \beta$, for every $c \in D(\beta)$, $\iota_\beta(A(c)) = \top$; $\iota_\alpha(\forall\, x.\, A(x)) = \bot$ iff there is $c \in D(\alpha)$ such that $\iota_\alpha(A(c)) = \bot$.

- $\iota_\alpha(\exists\, x.\, A(x)) = \top$ iff there exists a $c \in D(\alpha)$ such that $\iota_\alpha(A(c)) = \top$; on the contrary, $\iota_\alpha(\exists\, x.\, A(x)) = \bot$ iff, for every $\beta \in P$, $\alpha \leq \beta$, for every $c \in D(\beta)$, $\iota_\beta(A(c)) = \bot$.

One can easily verify that, for every formula A,

- if $\alpha \leq \beta$, and $\iota_\alpha(A) \neq \uparrow$, then $\iota_\alpha(A) = \iota_\beta(A)$.

- for every $\beta \in P$, β final, $\iota_\beta(A) \neq \uparrow$.

Comparing the Kripke models of the **E** logic with the corresponding models of the **IL** logic, one notices that

- the **E** logic has an explicit falsity in the evaluation function in order to build a constructible ¬ operator;

- in the **E** logic, the ¬ operator acts locally, i.e., its evaluation in a point of the frame does not depends on other points in the frame, while, in the **IL** logic, the evaluation of the ¬ operator is not local;

- the **E** logic requires the presence of final points in the frame, and they are used to define the semantics of the □ operator; on the contrary, the **IL** logic does not require final states.

It is easy to show that the evaluation function of **IL** acting on a final state of some **IL**-Kripke model coincides with the classical (Tarski) evaluation function. Hence, requiring final states allows to capture classical truth, since the classically valid formulae are exactly those valid in every final state.

Unfortunately, if one forces **IL**-Kripke model to have final states, as we did in **E**-Kripke models, the **IL** calculus is no more complete. In fact, **IL**-Kripke models with final states characterise the Kuroda logic, whose calculus is the same as **IL** plus

$$\frac{\forall x. \neg\neg A(x)}{\neg\neg \forall x. A(x)} \; \text{Kur}$$

The Kuroda calculus is sound and complete with respect to **IL**-Kripke models with final states, and it allows one to prove $\forall x. \neg\neg A(x) \rightarrow \neg\neg \forall x. A(x)$, that is not valid in **IL**.

The notion of truth in **E** is the following:

Definition 2.2.1 *A formula A is* valid *in the model $\mathfrak{K}$, notation $\vDash_{\mathfrak{K}} A$, iff, for every $\alpha \in P$, $\iota_\alpha(A) = \top$; a formula is* valid *(in **E**), notation $\vDash A$ iff it is valid in every model.*

A formula A is a consequence *of a set of formulae Γ in the model $\mathfrak{K}$, notation $\Gamma \vDash_{\mathfrak{K}} A$, iff for every $\gamma \in \Gamma$, such that $\vDash_{\mathfrak{K}} \gamma$, $\vDash_{\mathfrak{K}} A$; a formula A is a* consequence *of Γ (in **E**) iff, for every model $\mathfrak{K}$, $\Gamma \vDash_{\mathfrak{K}} A$.*

The Soundness Theorem states that, whenever A is deduced by Γ in the **E** calculus, then it is a consequence of Γ:

Theorem 2.2.1 (Soundness) *If $\Gamma \vdash A$ then $\Gamma \vDash A$.*

The Completeness Theorem states that, whenever the formula A is a consequence of Γ, A is provable from Γ:

Theorem 2.2.2 (Completeness) *If $\Gamma \vDash A$ then $\Gamma \vdash A$.*

The soundness and completeness of the **E** calculus have been proved in [38]. Since these proofs have no special interest in this work, we will omit them.

Proof-theoretical results

Some interesting facts are provable in the **E** logic:

Proposition 2.2.1 *The following facts are true in* **E***:*

1. $A \leftrightarrow \neg\neg A$;

2. $A \wedge B \leftrightarrow \neg(\neg A \vee \neg B)$;

3. $\neg(A \wedge B) \leftrightarrow \neg A \vee \neg B$;

4. $A \vee B \leftrightarrow \neg(\neg A \wedge \neg B)$;

5. $\neg(A \vee B) \leftrightarrow \neg A \wedge \neg B$;

6. $A \rightarrow \Box A$;

7. $\neg A \rightarrow \neg\Box A$;

8. $\Box A \leftrightarrow \Box\Box A$;

9. $\neg\Box A \leftrightarrow \Box\neg A$.

The first fact shows that negation in **E** is different from the negation in **IL**, since $\neg\neg A \rightarrow A$ is not valid in **IL**.

The facts 2 to 5 are the De Morgan's laws, which are not all valid in **IL**. Thus, their validity shows another significant difference in the interpretation of negation. Since De Morgan's laws are valid in **CL**, the negation of **E**, which is also called Nelson's negation from the name of its discoverer, is closer to the classical notion than to the intuitionistic one.

Facts 6 and 7 show that **E** is classically compatible, i.e., every formula A valid in **E** is also valid in **CL**. In fact, the following result shows that the interpretation of the $\Box$ operator as representing classical validity, is sound:

Theorem 2.2.3 $\Gamma \vdash_{\mathbf{CL}} A$ *iff* $\Gamma \vdash_{\mathbf{E}} \Box A$.

Proof: Let $\mathbf{E}_c$ be the E calculus plus the inference rule

$$\frac{\Box A}{A} \,\Box\mathrm{E} \ .$$

The $\mathbf{E}_c$ calculus is equivalent to **CL**, classical logic, since

$$\frac{\displaystyle \frac{\overset{[\neg A]}{\vdots}}{B} \quad \frac{\overset{[\neg A]}{\vdots}}{\neg B}}{A}\,\neg\mathrm{E}_{\mathbf{CL}} \quad = \quad \frac{\dfrac{\dfrac{\overset{[\neg A]}{\vdots}}{B} \quad \dfrac{\overset{[\neg A]}{\vdots}}{\neg B}}{\Box A}\,\Box\mathrm{I}}{A}\,\Box\mathrm{E} \quad .$$

Thus, $\Gamma \vdash_{\mathbf{E}} \Box A$ implies $\Gamma \vdash_{\mathbf{E}_c} \Box A$, hence $\Gamma \vdash_{\mathbf{CL}} A$.

Vice versa, if $\Gamma \vdash_{\mathbf{CL}} A$ then $\Gamma \vdash_{\mathbf{E}_c} A$; since $\vdash_{\mathbf{E}} B \to \Box B$ for every formula B, it suffices to prove that $\{\Box\gamma \,|\, \gamma \in \Gamma\} \vdash_{\mathbf{E}} \Box A$. This is done by structural induction on the proof in $\mathbf{E}_c$. $\qquad\square$

An important negative property of the **E** logic is that it does not obey to the principle of replacement of equivalents: if $\Gamma \vdash_{\mathbf{E}} A(P)$ where $A(P)$ is a formula with the P subformula in evidence, and we know that $P \leftrightarrow Q$, then it does not follow that $\Gamma \vdash_{\mathbf{E}} A(Q)$. In fact, $\neg(\Box\neg A \land \Box\neg B) \leftrightarrow \Box A \lor \Box B$ and $\Box\neg(A \lor B) \leftrightarrow \Box\neg A \land \Box\neg B$ are valid in **E**, but replacing the latter equivalence in the former gives $\Box(A \lor B) \leftrightarrow \Box A \lor \Box B$, that is not valid, as it is easy to see by a counter-model.

It is interesting to compare the deductive power of **IL** and **E**. On the propositional fragment, the following result holds

Theorem 2.2.4 *There exists a translation τ from* **IL** *formulae to* **E** *formulae such that, if $\vdash_{\mathbf{IL}} A$ then $\vdash_{\mathbf{E}} \tau(A)$, and vice versa.*

Proof: Let us define $\tau(p) = p$ for any atom, and $\tau(\neg A) = \neg\Box(\tau(A))$, $\tau(A \land B) = \tau(A) \land \tau(B)$, $\tau(A \lor B) = \tau(A) \lor \tau(B)$ and $\tau(A \to B) = \tau(A) \to \tau(B)$. The details of the proof can be found in [38]. $\qquad\square$

The translation from **IL** to **E** shows that, on the propositional fragment, $\neg_{\mathbf{IL}} \equiv \neg\Box$. On the contrary, **IL** cannot be embedded in **E** in the predicative setting. A partial justification of this fact can be given noticing that $\neg\Box \equiv \Box\neg$ in **E**, thus $\neg_{\mathbf{IL}} \equiv \Box\neg$. But negation in the scope of the $\Box$ operator has a classical meaning, and it is a well-known fact that $\vdash_{IL} \neg A$ is equivalent to $\vdash_{CL} \neg A$ only on the propositional fragment.

We conclude this section noticing that, according to [38], the **E** logic does not satisfy a normalisation theorem, in contrast with intuitionistic logic.

Chapter 3

Reasoning about Data

In this chapter we will speak about *specification frameworks* as a way to model the world where the verification, analysis and synthesis tasks take place. The notion of specification framework has been introduced in the context of Logic Programming and, now, it has a consolidated tradition [18, 19, 25].

The notion of specification framework is used to formalise the theories we use to reason about data; this approach gives raise to a highly uniform formal system that permits to verify programs, to analyse their correctness proofs and to synthesise programs from specification using a very homogeneous set of common concepts. In fact, as we will show in Chapter 5, the analysis of correctness proofs in our paradigm is the constructive basis that makes appealing the specification frameworks as a modelling technique.

Finding a solution to a problem implies an analysis phase where one constructs a language to describe the world the problem is posed in. Then one uses this language to state the properties that are supposed to be relevant to solve the problem itself. Finally, one writes down the specifications; in our approach every phase of this modelling process is formalised using the formal apparatus we are going to introduce.

More specifically, the first two steps correspond to the definition of a specification framework, while the last step is realised by writing the specifications in the language of the framework developed in the first two steps. Then, one writes the program implementing the specification, and, eventually, proves it correctness. Alternatively, one may *synthesise* a formally correct program from the specification.

In a verification task, we start from a program, a specification, and a context where the program exploits its action, we model these components in a constructive logical system, described as a specification framework in the **E** logic, we perform the verification, and, at the end, we are able to analyse the resulting proof in a formal, automatic way.

In a synthesis task, we start from a specification and a context, we model these components exactly in the same way as for the verification task, we proceed developing a suitable proof for the specification, and, then, we extract a program from the proof that is guaranteed to be formally correct.

3.1 Specification frameworks

The logical language we will adopt is the one of multi-sorted first-order logic [5]. In the following we assume the standard terminology and notations. Before starting with the definitions, we want to remark that both **IL** and **E** are multi-sorted first-order logics, thus we will treat them together.

A *signature* is a set of constants, function symbols and relational symbols with a specified arity. A Σ-structure $\mathfrak{S}$, that is, a (classical) model on the signature Σ, is Σ-*reachable* if every element in the domains can be denoted by a closed term. In the theory of abstract datatypes, this property is also referred to as *no junk property* [23].

In this context, a *specification framework* $\mathbb{F} = \langle \Sigma, \mathrm{Ax} \rangle$ is composed of a signature Σ, and a finite or recursive set Ax of Σ-axioms [25]. We distinguish between *closed* and *open* (specification) frameworks. This distinction is formalised using *isoinitial models*; this concept permits to single out the intended models a framework is supposed to speak of. A formal treatment of isoinitial models is given in [33, 36].

Definition 3.1.1 *Let T be a theory and let $\mathfrak{M}$ be a classical model for T. We say that $\mathfrak{M}$ is an* isoinitial *model for T iff, for every model $\mathfrak{N}$ of T, there is a unique isomorphic embedding from $\mathfrak{M}$ into $\mathfrak{N}$.*

It is immediate to show that the isoinitial model of a theory, if it exists, is unique up to isomorphisms.

Intuitively, an isoinitial model captures the notion of *minimal* model of a theory T with respect to isomorphic embeddings, that is, the isoinitial model is the unique model that is a submodel of every model of T. In other words, the isoinitial model is the kernel that is present in every model of T.

It is easy to see that there are theories admitting an isoinitial model, e.g., Peano arithmetic, where the standard model of natural numbers is isoinitial, and that there are theories not admitting an isoinitial model, e.g., the theory of groups.

Moreover, restricting the attention to reachable theories is not a limitation in the applicative context of Computer Science, since every element we may want to compute on, must be represented in some way in the computer memory, thus it must be representable in the formal language of a theory rich enough to model the computations.

3.2 Closed specification frameworks

The notion of closed framework formalises the datatypes that are commonly referred to as *concrete*, a superset of the *scalar* datatypes, opposed to the notion of *abstract* datatypes.

Definition 3.2.1 (Closed Framework) *Let* $\mathbb{F} = \langle \Sigma, Ax \rangle$ *be some specification framework; it is* closed *iff there is a Σ-reachable isoinitial model $\mathfrak{M}$ for Ax. We call $\mathfrak{M}$ the* intended model *of Ax.*

Thus, our closed frameworks are *isoinitial theories*, i.e., theories with a reachable isoinitial model. They are similar to *initial theories*, that axiomatise reachable initial models. The latter are quite popular in algebraic abstract datatypes and specifications [22, 24]. The difference is that initial models are based on homomorphisms, instead of isomorphic embeddings.

The notion of isoinitial theory, i.e., a theory admitting a reachable isoinitial model, can be characterised in a syntactical way, instead of a model-theoretic notion. In [31], the following condition has been shown:

Definition 3.2.2 *A theory T is said to be* atomically complete *iff, for every closed atomic formula ϕ, $T \vdash_{\text{CL}} \phi$ or $T \vdash_{\text{CL}} \neg\phi$.*

Theorem 3.2.1 *A theory T has a reachable isoinitial model iff T has a reachable model and it is atomically complete.*

Proof: Let $\mathfrak{M}$ be a reachable isoinitial model of T; by contradiction, let $T \nvdash_{\mathbf{CL}} r(t_1, \ldots, t_n)$ and $T \nvdash_{\mathbf{CL}} \neg r(t_1, \ldots, t_n)$ for some closed atomic formula $r(t_1, \ldots, t_n)$ in the language of T.

Then $T \cup \{r(t_1, \ldots, t_n)\}$ and $T \cup \{\neg r(t_1, \ldots, t_n)\}$ are consistent, thus there are the models $\mathfrak{N}_1$ and $\mathfrak{N}_2$ such that $\mathfrak{N}_1 \vDash T \cup \{r(t_1, \ldots, t_n)\}$ and $\mathfrak{N}_2 \vDash T \cup \{\neg r(t_1, \ldots, t_n)\}$.

Being isoinitial, we can isomorphically embed $\mathfrak{M}$ into $\mathfrak{N}_1$ and into $\mathfrak{N}_2$, hence $\mathfrak{N}_1 \vDash \neg r(t_1, \ldots, t_n)$ or $\mathfrak{N}_2 \vDash r(t_1, \ldots, t_n)$, depending whether $\mathfrak{M} \vDash \neg r(t_1, \ldots, t_n)$ or $\mathfrak{M} \vDash r(t_1, \ldots, t_n)$, respectively. In both cases, we get a contradiction, thus T must be atomically complete.

Vice versa, let $\mathfrak{M}$ be a reachable model of T and let T be atomically complete; let us call i the interpretation of the symbols of the language of T in $\mathfrak{M}$. It is immediate to see that $i(r) = \{\langle i(t_1), \ldots, i(t_n) \rangle \mid T \vdash_{\mathbf{CL}} r(t_1, \ldots, t_n)\}$ since, for every $\langle e_1, \ldots, e_n \rangle \in i(r)$, there are the terms $t_1, \ldots, t_n$ such that $\langle e_1, \ldots, e_n \rangle = \langle i(t_1), \ldots, i(t_n) \rangle$, by reachability of $\mathfrak{M}$, and moreover, if $\mathfrak{M} \vDash r(t_1, \ldots t_n)$ then $T \nvdash_{\mathbf{CL}} \neg r(t_1, \ldots, t_n)$, thus $T \vdash_{\mathbf{CL}} r(t_1, \ldots, t_n)$.

Let $\mathfrak{N} \vDash T$ be a model of T and let $\mathfrak{N}'$ be the model whose universe is the set of elements of $\mathfrak{N}$ denoted by some term, and whose interpretation j' is the restriction of the interpretation j of $\mathfrak{N}$ to the universe of $\mathfrak{N}'$. It is a standard result in model theory, see, e.g., [9], that $\mathfrak{N}'$ is a submodel of $\mathfrak{N}$, thus $\mathfrak{N}' \vDash T$.

Our claim is that $\mathfrak{M}$ is isomorphic to $\mathfrak{N}'$ by means of the map $g\colon i(t) \mapsto j'(t)$. In fact, g is a function, being $\mathfrak{M}$ reachable; it is obviously surjective; it is injective, since if $i(t_1) \neq i(t_2)$ then $\mathfrak{M} \vDash \neg t_1 = t_2$, thus $T \vdash \neg t_1 = t_2$ by atomic completeness, hence $\mathfrak{N}' \vDash \neg t_1 = t_2$, that is $j'(t_1) \neq j'(t_2)$.

It is immediate to show that g is a morphism between $\mathfrak{M}$ and $\mathfrak{N}'$, thus $\mathfrak{M}$ is isomorphic to $\mathfrak{N}'$, hence $\mathfrak{M}$ can be isomorphically embedded into $\mathfrak{N}$. □

Let us suppose that T is **E**-constructive[1] (**IL**-constructive), then the test for atomic completeness is reduced to prove that $\forall x. r(x) \lor \neg r(x)$ for any relation symbol r of the signature. This formula is obviously true in **CL**, classical logic, but not in

---

[1] In general, a theory T is L-constructive when $T + L$ is constructive.

a constructive logic; whenever we are able to prove it, we can immediately deduce that, for every term t, either $r(t)$ holds, or $\neg r(t)$ holds, as soon as the logical system is classically compatible, i.e., every deduction in the constructive system is a legal deduction in the classical system[2]. This is formalised in the following proposition:

Proposition 3.2.1 *If, for every relational symbol r of arity n in the signature of the* **E**-*constructive* (**IL**-*constructive*) *theory T, one can prove in* **E** (**IL**, *respectively*) *that*

$$\forall x_1, \ldots, x_n. r(x_1, \ldots, x_n) \vee \neg r(x_1, \ldots, x_n) \; ,$$

then T is atomically complete.

To ensure that a theory has a reachable model we adopt a prescriptive approach, requiring a condition on datatypes:

Proposition 3.2.2 *For every finite set of constructors C, i.e., constants and function symbols, and for every congruence relation $\approx$, the quotient via $\approx$ of the term algebra generated by C forms a reachable model over the signature $C \cup \{=\}$.*

Proof: The set of terms generated by C is reachable by construction, hence the quotient is reachable as well. Because of the interpretation of equality, two terms denote the same element iff they are congruent via $\approx$, hence the quotient is a model of the given signature. □

Hence, when defining a closed specification framework, we have to provide:

1. a signature Σ, containing one or more types $\tau_1, \ldots, \tau_n$, some constants and function symbols on these types, and some relation symbols on these types;

2. a special set of axioms, Eq formalising the congruence relation of Proposition 3.2.2;

3. a set of axioms A that forms the main body of the specification framework; they, together with Eq should form a constructive theory.

---

[2]This notion of compatibility is very syntactical, requiring only the repeatability of proofs in the classical system: the classical interpretation of the constructive system may even be inconsistent.

The logical theory behind the closed specification framework as well as its intended model are captured by the following result [31]:

Theorem 3.2.2 *For every set C of constructors, i.e., constant and function symbols, the term algebra generated by C is an isoinitial model for the theory*

$$T(C) = Identity\ theory + injectivity\ axioms + \\ + structural\ induction\ principle\ .$$

*Moreover, $T(C)$ is E-constructive (**IL**-constructive).*

The proof of Theorem 3.2.2 is rather obvious since the term algebra $\mathfrak{M}$ is a model of $T(C)$ by construction and, being a term model, it is reachable, and the only relation symbol is $=$, thus it is atomically complete by injectivity axioms. The proof that $T(C)$ is E-constructive (**IL**-constructive) is a direct consequence of the results in Chapter 5.

Usually, we are interested in extending an existing framework constructed as above, adding new axioms, or enlarging the signature. In these cases, we want to have an extension mechanism not changing the intended model except for the new symbols we may define, i.e., we want to make model-preserving *extensions*.

Definition 3.2.3 *Let $S = \langle \Sigma, A \rangle$ be a closed framework, let Σ' be a signature containing Σ, and let A' be a theory on the signature Σ' such that $A \subseteq A'$. Then $S' = \langle \Sigma', A' \rangle$ is an extension of S iff S' is a closed framework such that, being $\mathfrak{M}$ the intended model of S, the intended model $\mathfrak{M}'$ of S' is an expansion of $\mathfrak{M}$, i.e., the interpretation of Σ in $\mathfrak{M}'$ equals the interpretation of Σ in $\mathfrak{M}$.*

When we want to extend a closed framework by a new constant or function symbol, the following result applies:

Lemma 3.2.1 *Let $S = \langle \Sigma, A \rangle$ be a closed specification framework in the **E** logic, let A be E-constructive, and let*

$$A \vdash_E \forall x_1, \ldots, x_n. \exists! y. F(x_1, \ldots, x_n, y)\ ,$$

then, being f a new function symbol,

$$\langle \Sigma \cup \{f\}, A \cup \{\forall x_1, \ldots, x_n. F(x_1, \ldots, x_n, f(x_1, \ldots, x_n))\} \rangle$$

is an extension of S.

Analogously, we treat the addition of a relation symbol:

Lemma 3.2.2 *Let* $S = \langle \Sigma, A \rangle$ *be a closed specification framework in the* **E** *logic, let* A *be* **E**-*constructive, and let*

$$A \vdash_{\mathbf{E}} \forall\, x_1, \ldots, x_n.\, H(x_1, \ldots, x_n) \vee \neg H(x_1, \ldots, x_n) \ ,$$

then, being r *a new relation symbol*

$$\langle \Sigma \cup \{r\}, A \cup \{\forall\, x_1, \ldots, x_n.\, r(x_1, \ldots, x_n) \leftrightarrow H(x_1, \ldots, x_n)\}\rangle$$

is an extension of S.

Similar results hold in **IL** logic.

The other possibility when extending a closed specification framework, is to add new axioms. Let $S = \langle \Sigma, A \rangle$ be a closed specification framework and let B be a set of axioms on the signature Σ, we want a mechanism that ensures that $\langle \Sigma, A \cup B \rangle$ is a closed specification framework.

Our proposal is that, if S is an **IL**-framework, then B has to be an Harrop theory, and, that, if S is an **E**-framework, then B has to be a $\Box$-theory.

Formally, a $\Box$-theory is a set of **E**-Harrop formulae, and an **E**-Harrop formula is defined as follows

Definition 3.2.4 *A formula* ϕ *is said to be an* **E**-*Harrop iff*

- ϕ *is atomic or negated atomic;*

- $\phi \equiv \Box \psi$ *or* $\phi \equiv \neg \Box \psi$;

- $\phi \equiv \psi \wedge \theta$ *and both* ψ *and* θ *are* **E**-*Harrop formulae;*

- $\phi \equiv \psi \rightarrow \theta$ *and* θ *is an* **E**-*Harrop formula;*

- $\phi \equiv \forall\, x.\, \psi$ *and* ψ *is an* **E**-*Harrop formula.*

analogously, an Harrop formula is defined as

Definition 3.2.5 *A formula* ϕ *in the* **IL** *language is said to be an Harrop formula iff*

- ϕ *is atomic or negated atomic;*

- $\phi \equiv \psi \wedge \theta$ *and both* ψ *and* θ *are Harrop formulae;*

- $\phi \equiv \psi \rightarrow \theta$ *and* θ *is an Harrop formula;*

Framework PA
SORTS : $\mathbb{N}$;
FUNCTIONS : $0 : [] \to \mathbb{N}$;
                    $s : [\mathbb{N}] \to \mathbb{N}$;
                    $+, \cdot : [\mathbb{N}, \mathbb{N}] \to \mathbb{N}$;
RELATIONS : $=: [\mathbb{N}, \mathbb{N}]$
AXIOMS : $\forall x. \neg 0 = s(x)$; $\forall x, y. s(x) = s(y) \to x = y$;
                    $\forall x. x + 0 = x$; $\forall x, y. x + s(y) = s(x + y)$;
                    $\forall x. x \cdot 0 = 0$; $\forall x, y. x \cdot s(y) = x + x \cdot y$;
                    $H(0) \wedge \forall x. (H(x) \to H(s(x))) \to \forall x. H(x)$

Figure 3.1: The closed specification framework **PA**.

- $\phi \equiv \forall x. \psi$ *and* ψ *is an Harrop formula.*

Our proposal is supported by the fact that $T + E$, where T is a $\square$-theory, is constructive, as proved in Chapter 5, and by the fact that the isoinitial model is not changed, provided that T is E-consistent. Analogous results hold for the **IL** logic, with respect to Harrop theories.

There are other kind of admissible axioms, for example induction principles not generated via the type construction mechanism, like the descending chain principle we will discuss later. These axioms preserve both the constructive character of the theory and the reachability of the model.

To conclude this section, we want to show that the **IL** and **E** interpretation of arithmetic is, indeed, an example of a closed specification framework.

The closed specification framework for Peano arithmetic is shown in Figure 3.1. It is the result of a construction that goes as follows:

- First, we build the a closed framework **PA$_0$**, including the theory of identity, as follows

Framework PA$_0$
SORTS : $\mathbb{N}$;
FUNCTIONS : $0 : [] \to \mathbb{N}$;
                    $s : [\mathbb{N}] \to \mathbb{N}$;
RELATIONS : $=: [\mathbb{N}, \mathbb{N}]$

since the intended model of **PA$_0$** is the term algebra generated by the constructors 0 and s, we are guaranteed that this is a closed specification framework. Thus, we immediately synthesise the type $\mathbb{N}$ denoting the term algebra and some additional axioms characterising it:

$$\forall x. \neg 0 = \mathsf{s}(x)$$
$$\forall x, y. \mathsf{s}(x) = \mathsf{s}(y) \rightarrow x = y$$
$$H(0) \wedge \forall x. (H(x) \rightarrow H(\mathsf{s}(x))) \rightarrow \forall x. H(x)$$

The first two axioms are the so-called injectivity axioms, while the third family of axioms is the structural induction schema.

- Then, we can extend $\mathbf{PA_0}$ to $\mathbf{PA}$, by enlarging the signature with two new function symbols, $+$ and $\cdot$, and adding a series of axioms describing their behaviour

$$\forall x. x + 0 = x$$
$$\forall x, y. x + \mathsf{s}(y) = \mathsf{s}(x + y)$$
$$\forall x. x \cdot 0 = 0$$
$$\forall x, y. x \cdot \mathsf{s}(y) = x + x \cdot y$$

Being **E-Harrop** (Harrop, respectively) formulae these axioms are acceptable, if we are able to prove atomic completeness, that reduces to prove the formula

$$\forall x, y. x = y \vee \neg x = y \ .$$

Since this formula can be easily proved by induction, these axioms form a model-preserving extension of the isoinitial model for $\mathbf{PA_0}$.

3.3   Open specification frameworks

An open specification framework generalises the notion of closed framework. In particular, an open framework models a class of homogeneous closed frameworks. In this sense, an open framework represents an abstract datatype, like, e.g., lists, while a closed framework instancing the open framework, represents a concrete instance of the abstract datatype, e.g., lists of integers, of characters, of lists of integers, ...

Differently from a closed framework, an open framework depends on some parameters and defines a class of isoinitial theories. A *parametric signature* is a signature $\Sigma(P)$ where some symbols, the ones occurring in the list P, are put into evidence as *parameters*, see, e.g., [33]. A *parametric theory* $\text{Th}(P)$ over $\Sigma(P)$ is any $\Sigma(P)$-theory.

We can write $\text{Th}(P) = \mathbf{C}_P \cup \text{Ax}$, where $\mathbf{C}_P$ is the set of *constraints*, that is, axioms containing *only* parametric symbols and Ax is the set of *internal axioms*, containing at least a non-parametric symbol. The internal axioms are intended to formalise the defined symbols of $\Sigma(P)$, while the constraints represent requirements to be satisfied by actual parameters.

Definition 3.3.1 (Open Framework) *Let P be a set of parameters, let $\Sigma(P)$ be a parametric signature and let $T(P) = C_P \cup A$ be a parametric theory. The structure $\mathbb{F}(P) = \langle \Sigma(P), T(P) \rangle$ is an* open specification framework *iff, for every closed framework $\mathbb{C} = \langle \Sigma_C, A_C \rangle$ such that $P \subseteq \Sigma_C$ and $\mathbb{C} \vdash C_P$, $\mathbb{F}(\mathbb{C}) = \langle \Sigma_C \cup \Sigma(P), A_C \cup A \rangle$ is a closed specification framework. We call instance of $\mathbb{F}(P)$ with $\mathbb{C}$, the closed framework $\mathbb{F}(\mathbb{C})$. The* intended models *of $\mathbb{F}(P)$ are the intended models of all its instances.*

One may change the definition by noticing that it is not necessary to require $P \subseteq \Sigma_C$, but asking for something less, in particular, that there is a signature morphism preserving the parameters P, details can be found in [33]. In the following we adopt the above definition for clarity.

We are assured that an instance of an open framework is a closed framework if the internal axioms form an Harrop theory in **IL**, or a $\Box$-theory in **E**, as immediately follows from the properties we introduced in the previous section.

When we use an open framework, we want to instantiate its parameters to have a closed framework where we inherit the whole set of theorems proved in the open framework.

As an example of open framework (Figure 3.2), we show a characterisation for lists; this presentation differs from the standard algebraic description of lists, but it has the advantage to model direct access to elements.

The intended models of **LIST**$(\text{Elem}, \lhd)$ are the usual list structures with a partial ordering $\lhd$ on the (parametric) element type. Natural numbers, the function $\text{nocc}(x, L)$ (number of occurrences of x in L) and $\text{nth}(L, i, a)$ (a occurs in L at position i) have been introduced in this framework to make possible to reason about lists as a structured aggregation of

Framework LIST(Elem, $\lhd$)

IMPORT:	**PA**;
SORTS:	$\mathbb{N}$;
	Elem;
	List;
FUNCTIONS:	nil : [] $\to$ List;
	. : [Elem, List] $\to$ List;
	nocc : [Elem, List] $\to$ $\mathbb{N}$;
RELATIONS:	nth : [List, $\mathbb{N}$, Elem];
	$\preceq$: [List, List];
	$\lhd$: [Elem, Elem];
AXIOMS:	$\forall\, a, B. \neg\texttt{nil} = a\,.\,B$;
	$\forall\, a_1, a_2, B_1, B_2.\, a_1\,.\,B_1 = a_2\,.\,B_2 \to a_1 = a_2 \wedge B_1 = B_2$;
	$H(\texttt{nil}) \wedge (\forall\, a, J.\, (H(J) \to H(a\,.\,J))) \to \forall\, L.\, H(L)$;
	$\forall\, a, b, L.\, a = b \to \texttt{nocc}(a, b\,.\,L) = \texttt{nocc}(a, L) + 1$;
	$\forall\, a, b, L.\, \neg a = b \to \texttt{nocc}(a, b\,.\,L) = \texttt{nocc}(a, L)$;
	$\forall\, a, b, L.\, \texttt{nth}(a\,.\,L, 0, b) \leftrightarrow a = b$;
	$\forall\, a, b, i, L.\, \texttt{nth}(b\,.\,L, \texttt{s}(i), a) \leftrightarrow \texttt{nth}(L, i, a)$;
	$\forall\, a, b, A, B.\, a \lhd b \wedge A \preceq B \leftrightarrow a\,.\,A \preceq b\,.\,B$;
	$\forall\, a, b, A, B.\, \neg a \lhd b \to \neg a\,.\,A \preceq b\,.\,B$;
CONSTRAINTS:	$\forall\, x, y.\, x \lhd y \vee \neg x \lhd y$
	$\forall\, x, y.\, x \lhd y \wedge y \lhd x \leftrightarrow x = y$;
	$\forall\, x, y, z.\, x \lhd y \wedge y \lhd z \to x \lhd z$.

Figure 3.2: The open framework for lists.

elements, and, having direct access to elements through the nth function, to make easier to write down specifications.

The first three axioms define a list in the standard way by means of the nil and . constructors: they form an inductive definition of lists and they follow the term algebra pattern. The other axioms are Harrop and E-Harrop formulae, thus they form an extension of the closed framework of elements that is used to generate an instance.

The first constraint makes sense, since the **E (IL)** system is constructive, thus this instance of the excluded middle principle is not guaranteed to hold. The two last constraints model the fact that $\lhd$ is a partial ordering on elements.

Chapter 4

Program Synthesis

In this chapter we want to introduce a computational reading for a subclass of formulae of the **E** language. The interpretation we want to show is based on the idea that a formula may be read as a specification, that is, as a declaration of a task to perform, using a specification framework.

For example, the formula $\exists z. z \cdot z = x \vee \Box(\neg \exists z. z \cdot z = x)$ in the context of the **PA** framework expresses, according to our interpretation, the task of deciding whether x is a perfect square or not, and in the former case it also expresses the task of computing the square root of x; while, in the same context, the specification $\Box(\exists z. z \cdot z = x) \vee \Box(\neg \exists z. z \cdot z = x)$ expresses the task of deciding whether x is a perfect square or not, without requiring the computation of the square root.

Thus, the ultimate goal of writing specifications is to interpret their proofs in the **E** system as programs that compute them. Hence, the computational reading of specification formulae becomes the specification of the program that is hidden in their proof, so the name. Moreover, in a way, the proof of the theorem which validates the specification becomes the correctness proof of the corresponding program. This fact also justifies the name of specification frameworks.

At last, programs themselves can be used to suggest what is the "shape" of their correctness proofs. In fact, following the control flow of the program, we can think to a program as a sort of proof schema that transforms a specification,

41

thought to as the conclusion of the proof, into the axioms of the logical theory the program operates on. But, the "proof schema" may have a number of holes, which stand for facts to be proved having a logical meaning but no computational influence. This view of programs as proof-schemata will be partially investigated in the development of this chapter.

4.1 Specifications

We consider specifications of the form $\Gamma \Rightarrow \phi$, where $\Rightarrow$ is meta-implication, and Γ and ϕ are formulae in the E language with no occurrences of $\forall$ and $\rightarrow$, except under the scope of a $\square$ operator, and with negations only on atomic formulae or inside boxed formulae.[1]

Our restriction permits to capture the behaviour of every program not involving higher-order features. Moreover, this restriction allows to simplify the formal understanding of specifications; in any case, the semantics of specifications as used here is perfectly compatible with the general semantics based on evaluation forms, see [38].

Just as a glimpse, the evaluation form associated with a formula is, more or less, the lambda term which represents the formula in the Curry-Howard isomorphism, or, if you prefer, in the proofs-as-programs paradigm [3, 21]. We don't want to go any further in this direction, since it has no direct application to the program synthesis techniques we will develop.

Reading a specification as a computational requirement, it states a correctness requisite on a program. Its precise meaning can be stated using the semantics of evaluation forms, explained in [38]. Here, we give this semantics in a simplified form, oriented to explain our computational interpretation of constructive proofs.

To this aim, we associate with every specification formula θ the set of its *free individual variables* Var_θ, which corresponds to the set of its free variables in the usual sense, and a set V_θ of its evaluation variables. An assignment of V_θ codifies an eval-

---

[1] This last requirement is not restrictive, since, having De Morgan's Laws, we can adopt the standard techniques to move negations "inside".

uation form of θ, i.e., a possible explanation of its truth [38].

We inductively define V_θ as follows:

- if θ is atomic, negated atomic or a $\square$-formula, V_θ is empty, since no explanation of its truth is required;

- if $\theta \equiv \alpha \lor \beta$, then $V_\theta = V_\alpha \cup V_\beta \cup \{tv_\theta\}$, where tv_θ is a new boolean variable; in an isoinitial model $\mathfrak{M}$ and an assignment J of Var_θ, its meaning is "if tv_θ is false then α is true else β is true";

- if $\theta \equiv \alpha \land \beta$, then $V_\theta = V_\alpha \cup V_\beta$; its meaning is recursively explained by the meaning of α and β;

- if $\theta \equiv \exists x. \alpha$, then $V_\theta = V_\alpha \cup \{x_\theta\}$, where x_θ is a new variable with the sort of x; in an isoinitial model $\mathfrak{M}$ and an assignment J of Var_θ, the meaning of $(x_\theta = t) \in J$ is "$\exists x. \alpha$ is true because α is true by assigning the value t to x".

For a set Γ of formulae, Var_Γ and V_Γ are defined as the unions of, respectively, Var_α and V_α, for $\alpha \in \Gamma$.

Now, in a closed framework $\mathbb{F}$ with isoinitial model $\mathfrak{M}$, a specification $\Gamma \Rightarrow \phi$ is interpreted as: for any assignment J of the individual sequent variables $Var_\Gamma \cup Var_\phi$ and for any assignment I of V_Γ, if I and J make Γ true in $\mathfrak{M}$, then we want to compute an assignment I' of V_ϕ, such that I' and J make ϕ true[2] in $\mathfrak{M}$.

For example, in the previously introduced **PA** framework, let us consider the sequent

$$\exists z. z + z = x \lor z + sz = x \Rightarrow \exists z. z + z = sx \lor z + sz = sx$$

There is just one sequent variable, x; let

$$V_{\exists z. z+z=x \lor z+sz=x} = \{z_1, tv_1\} \quad \text{and}$$
$$V_{\exists z. z+z=sx \lor z+sz=sx} = \{z_2, tv_2\}$$

An assignment that makes the antecedent Γ true is, e.g., $x = ss0$, $z_1 = s0$, $tv_1 = false$; the correct output assignment is $z_2 = s0$, $tv_2 = true$. So a correct procedure is the following:

---

[2]A formal definition of what we intend for "to make true" is omitted for conciseness. It follows in the obvious way from the definition of V_θ and is informally explained through an example.

if tv_1 **then** $z_2 := sz_1$
      else $z_2 := z_1$
$tv_2 := $ **not** tv_1

The meaning of the program with respect to the specification sequent $\Gamma \Rightarrow \phi$ is: supposing to have a procedure that correctly computes the formula γ for every $\gamma \in \Gamma$, then the program computes the specification formula ϕ.

In the specific example, assuming to know how to compute z_1 such that $z_1 + z_1 = x \vee z_1 + sz_1 = x$, the program ensures that $z_2 + z_2 = sx \vee z_2 + sz_2 = sx$, being x the input value and z_2 the output value. Moreover, the tv_1 and tv_2 variables indicate whether the first or the second disjunct has been used in the two sequent formulae.

In another way, the specification sequent $\Gamma \Rightarrow \phi$ is satisfied by a program whose output makes ϕ true, and whose input makes Γ true[3]. Thus, Γ is the set of *preconditions* while ϕ is the *postcondition* of the synthesised program.

We want to remark that correctness in an open framework requires correctness in all its closed instances; thus, the program for computing the evaluation of the output formula is an open program, i.e., it may contain holes and uninterpreted functions. The requirement is correct reusability: the open program is correct if, when it gets instantiated to a closed instance, it becomes a complete and correct program.

As the reader may easily verify, the **IL**-frameworks can be used instead of the **E**-frameworks without significant differences. But the reader should notice that the restriction that negation may appear only on atomic formulae becomes strict since De Morgan's laws do not hold in **IL**.

4.2   Schemata for program synthesis

In this section, we want to introduce the way to synthesise programs from proofs. Our approach is based on the so-called *proof schemata*, a special kind of inference rules, that are derived in the **E** system plus a specification framework. In our view, we start from a specification formula, as described in the previous section, and we prove it in a framework using

---

[3]In this sentence, "to make true" has to be interpreted in a non-standard way, taking into account the elements of V_ϕ and V_Γ as *witnesses*.

the available proof schemata; as a result the proof can be compiled into a program, that correctly realises the specification.

The intuitive idea behind a proof schema S_{proof} is that it represents, at the same time, both a derived inference rule in a framework expressed in the **E** logic, and a partially specified program, that is, a program schema. Using schemata when deducing a specification goal permits to extract from the resulting proof a program that correctly implements the starting specification. For a complete account, see [1]; here we will just recall the fundamental inference rule, *dischargeability*, that assures computational completeness, i.e., that every program may eventually be derived.

Since it is quite common[4] to consider specifications of the form $\Gamma \Rightarrow \phi \vee \Box \neg \phi$, we find convenient to introduce a shortening notation: $\Gamma \Rightarrow \lceil \phi \rceil$. Let us consider a closed framework $\mathbb{F} = \langle \Sigma, \mathrm{Th} \rangle$ with isoinitial model $\mathfrak{S}$, and a specification[5] of the form

$$\Delta(\underline{x}) \Rightarrow \lceil \exists z. \psi(\underline{x}, z) \rceil .$$

The computability of the specification implies that the set of all elements $\underline{a}$ satisfying $\Delta(\underline{x})$ in the isoinitial model $\mathfrak{S}$ can be divided into two sets,

$$D^+ = \{\underline{a} \mid \mathfrak{S} \models \Delta(\underline{x}/\underline{a}) \cup \{\exists z. \psi(\underline{x}/\underline{a}, z)\}\} ,$$

and

$$D^- = \{\underline{a} \mid \mathfrak{S} \models \Delta(\underline{x}/\underline{a}) \cup \{\Box \neg \exists z. \psi(\underline{x}/\underline{a}, z)\}\} .$$

Now, let us suppose that there exist $n + m$ sets of formulae $\Gamma_1^+(\underline{x}), \ldots, \Gamma_n^+(\underline{x})$ and $\Gamma_1^-(\underline{x}), \ldots, \Gamma_m^-(\underline{x})$ such that:

(F1) $\underline{a} \in D^+$ iff there is $\Gamma_i^+(\underline{x})$, $1 \leq i \leq n$, s.t. $\mathfrak{S} \models \Gamma_i^+(\underline{x}/\underline{a})$;

(F2) $\underline{a} \in D^-$ iff there is $\Gamma_j^-(\underline{x})$, $1 \leq j \leq m$, s.t. $\mathfrak{S} \models \Gamma_j^-(\underline{x}/\underline{a})$.

(F3) For $1 \leq i \leq n$, $\Delta(\underline{x}), \Gamma_i^+(\underline{x}) \vdash_{\mathrm{Th}+\mathbf{E}} \psi(\underline{x}, t(\underline{x}))$ for an appropriate term $t(\underline{x})$;

---

[4]In fact, a formula $\phi \vee \Box \neg \phi$ encodes the *if* ... *then* ... *else* schema: we compute ϕ and if it is true, the *then* branch holds, otherwise we know, without calculation, that $\neg \phi$ holds, thus we choose the *else* branch.

[5]An underlined variable like $\underline{x}$ stands for a tuple of variables $x_1, \ldots, x_k$.

(F4) For $1 \leq j \leq m$, $\Delta(\underline{x}), \Gamma_j^-(\underline{x}) \vdash_{\text{Th+E}} \Box \neg \exists z. \psi(\underline{x}, z)$.

If we are able to satisfy conditions (F1)-(F4), we have "reduced" the problem of solving the specification to the problem of deciding, given a possible input $\underline{a}$ satisfying the preconditions $\Delta(\underline{x})$, which set $\Gamma_i^{\pm}(\underline{x}/\underline{a})$ gets satisfied in the isoinitial model. Obviously, the problem has been "reduced" if the formulae occurring in these sets are "simpler" than the formula representing the specification. If we can state that the set whose elements are $\Gamma_i^{\pm}(\underline{x})$, is *dischargeable* then we are able to generate a proof for ϕ in the **E** system not depending on any $\Gamma_i^{\pm}(\underline{x})$. Technical details can be found in [1]. We remark that the computation that checks if a set is dischargeable can be used to construct that proof.

The dischargeability rule, which reduces to an iterate application of $\exists$E and $\lor$E rules, can be interpreted as a program pattern; every element of the sets $\Gamma_i^{\pm}(\underline{x})$ is a test or a function we should compute in a case analysis structure following the proof pattern. To illustrate how this rule practically works, let us consider an example. Let us suppose to work in the framework of total orderings, and let us define

$$\min(a, b, c, m) \equiv m \leq a \land m \leq b \land m \leq c \land$$
$$\land (m = a \lor m = b \lor m = c) \ ;$$

we want to synthesise a program satisfying the specification[6]

$$\theta \equiv \exists m. \min(x, y, z, m) \ ,$$

that is, a program to compute the minimum element in a set of three elements. Using the framework, it is not difficult to prove the following facts:

- $\{x \leq y, x \leq z\} \Rightarrow \min(x, y, z, x)$

- $\{x \leq y, \neg x \leq z\} \Rightarrow \min(x, y, z, z)$

- $\{\neg x \leq y, y \leq z\} \Rightarrow \min(x, y, z, y)$

- $\{\neg x \leq y, \neg y \leq z\} \Rightarrow \min(x, y, z, z)$

---

[6]The specification $\Gamma \Rightarrow \phi$ is simplified to ϕ when Γ is empty.

The family of sets whose members are the preconditions of these facts, constitutes a dischargeable set, so we can use it to construct the following proof, which is an instance of the abstract dischargeability rule,

$$
\begin{array}{cccc}
[x \leq y] & [x \leq y] & [\neg x \leq y] & [\neg x \leq y] \\
[x \leq z] & [\neg x \leq z] & [y \leq z] & [\neg y \leq z] \\
\vdots & \vdots & \vdots & \vdots
\end{array}
$$

$$
\begin{array}{ccccccc}
& \vdots & \min(x,y,z,x) & \min(x,y,z,z) & \vdots & \min(x,y,z,y) & \min(x,y,z,z) \\
\vdots & \lceil x \leq z \rceil & \theta & \theta & \lceil y \leq z \rceil & \theta & \theta \\
\lceil x \leq y \rceil & & \theta & & & \theta \\
& & & \theta
\end{array}
$$

The synthesised program schema is

if $x \leq y$ **then**
   if $x \leq z$ **then** $m := x$
        else $m := z$
else
   if $y \leq z$ **then** $m := y$
        else $m := z$

In general, proving a specification in a framework involves some inductive reasoning. In the following, we will study two induction principles, DESCENDING CHAIN and DIVIDE ET IMPERA, showing how they act as schemata, i.e., how they can be interpreted as proof patterns. While other principles are studied in [1], here our focus is on lifting computational structures, specifically loops, on the logical level.

Descending chain principle

The descending chain principle has been introduced in the context of program synthesis in [33, 40, 41] as the counterpart of *repeat ... until* loops. Here we describe this principle applied to a specification of the form $\Delta(\underline{x}) \Rightarrow \exists z. \psi(\underline{x}, z)$. The principle can be extended to other specification forms. The corresponding inference rule, called *DCH*, is:

$$
\begin{array}{cc}
\Delta(\underline{x}) & \Delta(\underline{x}), [A(\underline{x}, y)] \\
\vdots \, \pi_1 & \vdots \, \pi_2 \\
\exists z. A(\underline{x}, z) & (\exists z. A(\underline{x}, z) \wedge z \prec y) \vee \exists z. \psi(\underline{x}, z) \\
\hline
\multicolumn{2}{c}{\exists z. \psi(\underline{x}, z)}
\end{array} \text{DCH}
$$

where y (the *parameter* of the rule) does not occur free in ψ and in the other undischarged assumptions of the π_2 proof, so y acts as an eigenvariable.

For the principle to be valid, the framework $\mathbb{F}(P)$ must satisfy the condition that the relation symbol $\prec$ is interpreted in any intended model of $\mathbb{F}(P)$ as a well founded order relation.

The formula $A(\underline{x}, y)$ is called the *invariant* (of the loop) since it is the condition that holds whenever the loop is under execution, as is customary in program verification, see [11]. This induction principle corresponds to a **repeat-until** loop that computes over a decreasing sequence of values with respect to $\prec$, approximating the solution; the solution is reached at the end of the cycle. Hence, the program schema corresponding to the DCH rule is:

P_1;
repeat
  P_2;
until (**not** tv_θ);

where P_1 and P_2 are the sub-programs obtained by translating the proofs π_1 and π_2 respectively, and tv_θ is the variable storing the boolean value associated with the formula $\theta \equiv (\exists z. A(\underline{x}, z) \wedge z \prec y) \vee \exists z. \psi(\underline{x}, z)$, as defined in the interpretation of specifications. For a detailed discussion, see [1].

Divide et impera

A frequently used solving method consist of partitioning the input into smaller instances and then to solve the original problem combining the result obtained from the solutions of the smaller instances. This strategy is called *Divide et Impera*, and, in many cases, it permits to obtain efficient algorithms.

Considering a specification $\Delta(x) \Rightarrow \exists z. \phi(x, z)$, the *Divide et Impera* method is described as

1. If $dg(x) \leq C$ for a fixed value C, the solution can be directly computed;

2. If $dg(x) > C$, the generated algorithm has to perform the following steps

   a) x is partitioned into $y_1, \ldots, y_n$ such that, for all j, $dg(y_j) < dg(x)$;

b) the specifications $\Delta(y_j) \Rightarrow \exists z_j.\phi(y_j, z_j)$ are recursively solved;

c) The witnesses t_j, from the specifications solved in the previous step, are combined to obtain the solution for x.

where dg is a measure function (degree) on the input space to some ordered structure.

The proof schema encoding the previous pattern is

$$
\cfrac{
\begin{array}{ccc}
\cfrac{\Delta(x), [dg(x) \leq C] \\ \vdots\ \pi_1(x) \\ \exists z.\phi(x,z)}{} &
\cfrac{\Delta(x) \\ \vdots\ \pi_2(x) \\ \exists \overline{y}.\text{Part}(x,\overline{y})}{} &
\cfrac{\begin{array}{c}\Delta(x), [\text{Part}(x,\overline{y})], \\ [\exists z_1.\phi(y_1,z_1)], \\ \ldots, \\ [\exists z_n.\phi(y_n,z_n)]\end{array} \\ \vdots\ \pi_3(x,\overline{y}) \\ \exists z.\phi(x,z)}{}
\end{array}
}{\exists z.\phi(x,z)}\ \text{DI}
$$

The program corresponding to the schema can be summarised as follows:

Procedure $F(x)$
begin
    if $dg(x) \leq C$ **then return** $P_1(x)$
    else
        begin
            $P_2(x, y_1, \ldots, y_n)$;
            for $j = 1$ **to** n **do** $z_j := F(y_j)$;
            $P_3(x, z_1, \ldots, z_n)$;
            return z;
        end
end

where P_1, P_2 and P_3 are the programs synthesised from π_1, π_2 and π_3, respectively, and Part is the predicate that represents the input partition procedure. More precisely, the specification $\Delta(x) \Rightarrow \exists \overline{y}.\text{Part}(x,\overline{y})$, proved by the π_2 subproof, defines how to partition the input.

This proof schema is valid only in a specification framework $\mathbb{F} = \langle \Sigma, \text{Th} \rangle$ where it is possible to prove that

$$
\vdash_{\text{Th}+\mathbf{E}} \forall\, x, y_1, \ldots, y_n.\ \text{Part}(x, y_1, \ldots, y_n) \wedge dg(x) > C \rightarrow
$$
$$
\rightarrow \bigwedge_{1 \leq j \leq n} dg(y_j) < dg(x)
$$

As an example, let us consider the specification formula $\exists x. \text{minArray}(A, x)$ in the open framework that axiomatises arrays; it is a direct generalisation of $\textbf{List}(\text{Elem}, \lhd)$.

The minArray predicate is defined as

$$\text{minArray}(A, x) \equiv (\exists n. \text{nth}(A, n, x)) \wedge$$
$$\wedge \Box (\forall i. \exists y. \text{nth}(A, i, y) \rightarrow x \leq y) \ .$$

To compute this specification, we take advantage of the min predicate, already synthesised using the dischargeability rule. A way to encode this idea, is the following: let us fix $C = 3$ and $dg(A) = \text{size}(A)$; our partition strategy is to divide the array into three pieces of the same size, and it is immediate to write a PART predicate encoding this requirement.

Applying the *Divide et Impera* proof schema to these definitions, the proof π_3 is essentially the proof we gave for min in Section 4.2; the proof π_2 has to be derived from the framework; the proof π_1 is by cases on $dg(A)$. The resulting program schema is

Procedure minArray(A)
begin
    if size$(A) \leq 3$ **then return** $P_1(A)$
    else
        begin
            $P_2(A, A_1, A_2, A_3)$;
            for $j = 1$ **to** 3 **do** $z_j := \text{minArray}(A_j)$;
            $\min(z_1, z_2, z_3, z)$;
            return z;
        end
end

We observe that the preceding program uses the already synthesised min procedure, that the program is parametric in P_1 and P_2, and it is correct if P_1 (P_2, respectively) has a correctness proof that matches π_1 (π_2, respectively) in the schema.

Chapter 5

The Collection Method

It is evident that a proof contains many true facts that, together, concur to establish the truth of the conclusion. In a correctness proof, many of these facts are strictly related to the verified program.

Our conception of analysis is to extract the truth content from a proof. A general way to think a proof is to imagine it as a sequence of steps linking the hypothesis to the conclusion. During this process, a series of true facts must be established, and a set of implicit, but obvious, consequences are derived.

Informally, the whole set of true facts, either explicit or implicit, established in the proof development forms its truth content. This is not a definition, since we have not described what we mean as implicit information, and we have not fixed any system specifying what are the proof steps we accept.

The goal of this introduction is to provide the intuition behind the formal instruments we will develop. For this reason we are not very precise now, but we prefer to give the flavour of our analysis methodology. In this respect, we feel free not to choose any formal logic in this moment, but to discuss what aspects of this choice are relevant for our purposes and why.

Our analysis starts from a formal proof Π: every fact that is true because we can exhibit a subproof of Π for it, is part of the information content of Π. We can derive more facts by combining the subproofs of Π; again, their conclusions are part of the information content of Π. Every instance of the eigenvariables of a subproof of Π is again something that, intuitively, we have proved, hence it should be part of the information content of Π, as well.

The core of the *collection method* is to provide a family of algorithms, all based on the same structure, that construct a set I of formulae that is a subset of the information content of a proof, and, at the same time, it contains enough formulae to give a complete account of the reasons why the proof is true, in a constructive perspective.

The last sentence needs an explanation: a set of formulae can be *evaluated* with respect to itself, that is, if it contains a conjunction then it must contain both conjuncts, if it contains a disjunction it must contain at least one of the disjuncts, and so on (again, the formal details are worked out later).

The collection method constructs the minimal set, closed under evaluation, containing the whole set of facts that can be directly extracted from the initial proof, i.e., the facts that can be derived from a proof by looking at its subproofs.

As we have seen in the previous lectures, it is important, in the context of analysis of correctness proofs, to exploit the constructive character of the logical system, since deciding disjunctions and existential statement gives a way to use proofs as programs. In this view, the collection method can be thought as a logical machine executing proofs [30, 32, 36, 37, 39, 40, 41, 42].

5.1 Intuitionistic logic

In this part, we will show the collection method applied to intuitionistic logic. The schema we are going to investigate can be applied to most constructive logics and theories.

The formal system, i.e., the logical calculus has already been shown in Table 2.1 and it is based on the language whose connectives are $\neg$, $\wedge$, $\vee$, $\rightarrow$, $\forall$, $\exists$, a set of variables denoted by lowercase Roman letters, and a set of uninterpreted function and predicate symbols. The first notion we are interested in is

Definition 5.1.1 *We say that π is a* subproof *of π', notation $\pi \prec \pi'$, where π and π' are proofs in the* **IL** *calculus, iff either π and π' are identical, modulo α-conversion[1] [4], or $\pi \prec \pi''$ with π'' an*

---

[1]Id est, consistent renaming of eigenvariables.

immediate subproof of π', i.e., a premise of the last applied inference rule in π'.

Moreover, let π be a proof, we define

$$\text{Subproof}(\pi) = \{\pi' \mid \pi' \prec \pi\} \; ;$$

if $\mathcal{I}$ is a set of proofs, $\text{Subproof}(\mathcal{I}) = \bigcup_{\pi \in \mathcal{I}} \text{Subproof}(\pi)$.

The collection method is based on two operations:

- composing subproofs of a given set of proofs;

- instantiating eigenvariables of parametric proofs to generate new proofs.

The collection operator performs the first operation: if

$$
\begin{array}{ccc}
\Gamma & & \Delta, A \\
\vdots\, 1 & \text{and} & \vdots\, 2 \\
A & & B
\end{array}
$$

are subproofs of $\mathcal{I}$, a fixed set of proofs, then the proof

$$
\begin{array}{ccc}
 & & \Gamma \\
\Gamma, \Delta & & \vdots\, 1 \\
\vdots & \equiv & A \quad , \Delta \\
B & & \vdots\, 2 \\
 & & B
\end{array}
$$

is an element of the collection over $\mathcal{I}$.

The key idea is that the direct truth content of the set of proofs $\mathcal{I}$ is given by the conclusions of the collected proofs without undischarged assumptions.

Definition 5.1.2 *Let $\mathcal{I}$ be a set of proofs;*

$$\text{Coll}(\mathcal{I}) = \bigcup_{i \in \omega} \text{Coll}^i(\mathcal{I}) \; ,$$

where

$$\text{Coll}^0(\mathcal{I}) = \emptyset$$

$$\text{Coll}^{i+1}(\mathcal{I}) = \text{Coll}^i(\mathcal{I}) \cup \left\{ \begin{array}{c} \Gamma \\ \vdots \\ A \end{array} \middle| \begin{array}{c} \Gamma \\ \vdots \\ A \end{array} \in \text{Subproof}(\mathcal{I}) \land \right.$$

$$\left. \land\, \Gamma \subseteq \{\phi \mid \exists \Pi \in \text{Coll}^i(\mathcal{I}). \Pi \vdash \phi\} \right\} \; ,$$

and the notation $\Pi \vdash \phi$ means that the formula ϕ is the conclusion of the proof Π.

Definition 5.1.3 *Let $\mathcal{I}$ be a set of proofs,*

$$\mathrm{Inf}(\mathcal{I}) = \{\phi \mid \phi \text{ is the conclusion of a proof in } \mathrm{Coll}(\mathcal{I})\} \ .$$

We have to remark that the definition of Coll provides an abstract algorithm implementing this operator.

As we previously stated, the second main operation underlying the collection method, is the instantiation of eigenvariables. The idea is simple: if $\pi(p)$ is a proof depending on the eigenvariable p, p is free in the proof, hence $\pi(p := t)$, the proof obtained by substituting the term t for p, is a valid proof. We can enrich the initial set of proofs $\mathcal{I}$, by adding to it instances of its parametric subproofs.

We need a compromise, of course: we should decide what instances are useful to add, and what are superfluous. Our proposal is to add instances to $\mathcal{I}$ that do not enlarge the set of conclusions of proofs in $\mathrm{Coll}(\mathcal{I})$. Of course they will eventually enlarge the set of conclusions in $\mathrm{Coll}(\mathcal{I} \cup N)$, where N is the set of instances.

Definition 5.1.4 *Let $\mathcal{I}$ be a set of proofs, we define*

$$\forall I\text{-}Sub(\mathcal{I}) = \left\{ \begin{pmatrix} \Gamma \\ \vdots \\ A(p) \end{pmatrix} (p := t) \ \middle| \ \Gamma \cup \{A(t)\} \in \mathrm{Inf}(\mathcal{I}) \ \wedge \right.$$

$$\left. \wedge \quad \dfrac{\begin{matrix} \Gamma \\ \vdots \\ A(p) \end{matrix}}{\forall\, x.\, A(x)} \forall I \quad \in \mathrm{Coll}(\mathcal{I}) \right\} \ .$$

Definition 5.1.5 *Let $\mathcal{I}$ be a set of proofs, we define*

$$\exists E\text{-}Sub(\mathcal{I}) = \left\{ \begin{pmatrix} \Gamma, A(p) \\ \vdots \\ B \end{pmatrix} (p := t) \ \middle| \ \Gamma \cup \Delta \cup \{A(t)\} \in \mathrm{Inf}(\mathcal{I}) \right.$$

$$\left. \wedge \quad \dfrac{\begin{matrix} \Delta & \Gamma, [A(p)] \\ \vdots & \vdots \\ \exists\, x.\, A(x) & B \end{matrix}}{B} \exists E \quad \in \mathrm{Coll}(\mathcal{I}) \right\} \ .$$

Since the $\forall$I and the $\exists$E inference rules are the only parametric rules in **IL**, the previous definitions cover the whole set of possible instantiations we allow with our policy.

Definition 5.1.6 *Given a set $\mathcal{I}$ of proofs, we define the expansion operator as*

$$\mathrm{Exp}(\mathcal{I}) = \mathcal{I} \cup \forall\text{I-}Sub(\mathcal{I}) \cup \exists\text{E-}Sub(\mathcal{I}) \ .$$

The expansion operation can be iterated, producing more and more proofs to analyse; since the Exp operator is monotone with respect to set inclusion, it has a least fixed point with a standard characterisation.

The limit of this construction provides a set of proofs and the information we extract from it via the Inf opertaor is the relevant constructive part of the truth content of the initial set of proofs.

Definition 5.1.7 *Let $\mathcal{I}$ be a set of proofs,*

$$\mathrm{Exp}^0(\mathcal{I}) = \mathcal{I} \ ,$$
$$\mathrm{Exp}^{i+1}(\mathcal{I}) = \mathrm{Exp}(\mathrm{Exp}^i(\mathcal{I})) \ .$$

Its closure is

$$\mathrm{Exp}^*(\mathcal{I}) = \bigcup_{i \in \omega} \mathrm{Exp}^i(\mathcal{I}) \ .$$

Moreover, we define

$$\mathrm{Coll}^*(\mathcal{I}) = \bigcup_{i \in \omega} \mathrm{Coll}(\mathrm{Exp}^i(\mathcal{I}))$$

and

$$\mathrm{Inf}^*(\mathcal{I}) - \bigcup_{i \in \omega} \mathrm{Inf}(\mathrm{Exp}^i(\mathcal{I})) \ .$$

An algorithm implementing the collection method generates $\mathrm{Inf}^*(\mathcal{I})$, where $\mathcal{I}$ is the input, a finite set of proofs. It should be evident that, when $\mathcal{I}$ is finite, $\mathrm{Inf}^*(\mathcal{I})$ is recursively enumerable. It is possible to prove that, in general [39], $\mathrm{Inf}^*(\mathcal{I})$ is not recursive.

The role of the Coll* operator is to keep track of the *interesting* subproofs, while the $\mathrm{Inf}^*(\mathcal{I})$ computes the *relevant* part of the truth content of a set of proofs $\mathcal{I}$. The words *interesting* and *relevant* are strictly related to the computational meaning of $\mathcal{I}$. The interesting property of $\text{Inf}^*(\mathcal{I})$ is that it allows a constructive reading of the logic **IL**, as well as a computational reading of formulae (proofs later):

- $A \wedge B \in \text{Inf}^*(\mathcal{I})$ implies $A \in \text{Inf}^*(\mathcal{I})$ and $B \in \text{Inf}^*(\mathcal{I})$; if $A \wedge B$ is a specification, it means that a program computing $A \wedge B$ has to compute both A and B.

- $A \vee B \in \text{Inf}^*(\mathcal{I})$ implies $A \in \text{Inf}^*(\mathcal{I})$ or $B \in \text{Inf}^*(\mathcal{I})$; if $A \vee B$ is a specification then a program computing $A \vee B$ has to compute A or to compute B, The other way around, the computational interpretation of a disjunction as a decision procedure between two alternatives is sound with the constructive reading we adopted.

- $\exists x. A(x) \in \text{Inf}^*(\mathcal{I})$ implies that there is a term t such that $A(t) \in \text{Inf}^*(\mathcal{I})$; being $\exists x. A(x)$ a specification, we have to compute a term t that satisfies $A(t)$. We can find this witness in $\text{Inf}^*(\mathcal{I})$, thus, again, the constructive reading and the computational reading coincide.

IL is uniformly constructive

The starting point is to define the notion of pseudo-truth set. The intuitive meaning we want to induce into this definition, is that of a syntactically consistent set. We want that every formula in this set is a theorem, and the whole set, in a sense, provides an explanation for itself, following the semantics of the logic we are working on.

Definition 5.1.8 *A set $\mathcal{F}$ of formulae is a* pseudo-truth set *iff*

- $A \in \mathcal{F}$ *implies* $\vdash A$.

- $\neg A \in \mathcal{F}$ *implies that* $A \notin \mathcal{F}$.

- $A \vee B \in \mathcal{F}$ *implies* $A \in \mathcal{F}$ *or* $B \in \mathcal{F}$.

- $A \wedge B \in \mathcal{F}$ *implies* $A \in \mathcal{F}$ *and* $B \in \mathcal{F}$.

- $A \rightarrow B \in \mathcal{F}$ *and* $A \in \mathcal{F}$ *implies* $B \in \mathcal{F}$.

- $\exists x. A(x) \in \mathcal{F}$ *implies that there is a* t, *such that* $A(t) \in \mathcal{F}$.

The notion of pseudo-truth set is global, that is, it looks at the whole set of formulae in a single glance; the corresponding local notion is that of evaluation. A formula is evaluated in a set of formulae if it is *explained* by that set.

Definition 5.1.9 *Let A be a formula, it is evaluated in $\mathcal{F}$, a set of formulae, if and only if*

- $A \in \mathcal{F}$.

- A *is an atomic or a negated formula.*

- $A \equiv B \wedge C$ *and B and C are evaluated in $\mathcal{F}$.*

- $A \equiv B \vee C$ *and B is evaluated in $\mathcal{F}$, or C is evaluated in $\mathcal{F}$.*

- $A \equiv B \to C$ *and, if B is evaluated in $\mathcal{F}$, then also C is evaluated in $\mathcal{F}$.*

- $A \equiv \exists x. B(x)$ *and there is a term t such that $B(t)$ is evaluated in $\mathcal{F}$.*

- $A \equiv \forall x. B(x)$ *and, for all terms t such that $B(t) \in \mathcal{F}$, the formula $B(t)$ is evaluated in $\mathcal{F}$.*

The main goal of our proving effort is to show that the global notion and the local notion coincide in the case of $\mathrm{Inf}^*(\mathcal{I})$, for any set $\mathcal{I}$ of proofs.

In order to gain this result, we need two closure lemmata; the first one proves that our construction is closed under the membership relation; the second one proves that the construction is closed under evaluation.

Lemma 5.1.1 *Let $\mathcal{I}$ be a set of proofs; if $\begin{array}{c}\Gamma \\ \vdots \\ A\end{array}$ is a subproof of a proof in $\mathrm{Exp}^*(\mathcal{I})$, and $\Gamma \subseteq \mathrm{Inf}^*(\mathcal{I})$ then $A \in \mathrm{Inf}^*(\mathcal{I})$.*

Proof: Let $\Gamma = \{B_1, \ldots B_n\}$. From $\begin{array}{c}\Gamma \\ \vdots \\ A\end{array} \prec \mathrm{Exp}^*(\mathcal{I})$, there is an

index j such that $\begin{array}{c}\Gamma \\ \vdots \\ A\end{array} \prec \mathrm{Exp}^j(\mathcal{I})$ and, for all $i \geq j$, $\begin{array}{c}\Gamma \\ \vdots \\ A\end{array} \prec \mathrm{Exp}^i(\mathcal{I})$.

From $\{B_1, \ldots, B_n\} \subseteq \text{Coll}^*(\mathcal{I})$, there are indexes $i_1, \ldots, i_n$ such that $B_1 \in \text{Inf}(\text{Exp}^{i_1}(\mathcal{I})), \ldots, B_n \in \text{Inf}(\text{Exp}^{i_n}(\mathcal{I}))$.

Let k be the maximum in $j, i_1, \ldots, i_n$, then

$$\begin{array}{c} \Gamma \\ \vdots \\ A \end{array} \prec \text{Exp}^k(\mathcal{I}) \quad \text{and} \quad \Gamma \subseteq \text{Inf}(\text{Exp}^k(\mathcal{I})) ,$$

hence, by definition, $A \in \text{Inf}(\text{Exp}^k(\mathcal{I}))$, so $A \in \text{Inf}^*(\mathcal{I})$. $\quad\square$

Lemma 5.1.2 *Let $\mathcal{I}$ be a set of proofs, and let* $\begin{array}{c} \Gamma \\ \vdots \\ A \end{array} \prec \text{Exp}^*(\mathcal{I})$, *and let Γ be evaluated in* $\text{Inf}^*(\mathcal{I})$, *then A is evaluated in* $\text{Inf}^*(\mathcal{I})$.

Proof: By Lemma 5.1.1, we know that $A \in \text{Inf}^*(\mathcal{I})$.

We prove that A satisfies the other condition to be evaluated by induction on the structure of proofs.

- Assumption:

$$\begin{array}{c} \Gamma \\ \vdots \\ A \end{array} \equiv A$$

$A \in \Gamma$, thus, by hypothesis, it is evaluated in $\text{Inf}^*(\mathcal{I})$.

- $\neg$ Introduction, $\neg$ Elimination: being the conclusion an atomic or negated formula, by definition, it is evaluated in $\text{Inf}^*(\mathcal{I})$.

- $\wedge$ Introduction, $\wedge$ Elimination, $\vee$ Introduction, $\rightarrow$ Elimination, $\forall$ Elimination, $\exists$ Introduction: by induction hypothesis, the conclusions of the immediate subproofs are evaluated in $\text{Inf}^*(\mathcal{I})$, thus, by definition, the conclusion is evaluated in $\text{Inf}^*(\mathcal{I})$, as well.

- $\vee$ Elimination:

$$\begin{array}{c} \Gamma \\ \vdots \\ A \end{array} \equiv \dfrac{\begin{array}{ccc} \Gamma & \Gamma,[C] & \Gamma,[D] \\ \vdots & \vdots\,{}^1 & \vdots\,{}^2 \\ C \vee D & A & A \end{array}}{A}$$

by inductions hypothesis, $C \vee D$ is evaluated in $\text{Inf}^*(\mathcal{I})$, so C or D is evaluated in $\text{Inf}^*(\mathcal{I})$, too.

If C is evaluated, then from the induction hypothesis on

$$\begin{array}{c} \Gamma, C \\ \vdots_1 \\ A \end{array}$$

we can deduce that A is evaluated in $\mathrm{Inf}^*(\mathcal{I})$; if D is evaluated, then from the induction hypothesis on

$$\begin{array}{c} \Gamma, D \\ \vdots_2 \\ A \end{array}$$

we can deduce that A is evaluated. Hence, A is evaluated in $\mathrm{Inf}^*(\mathcal{I})$.

- $\to$ Introduction:

$$\begin{array}{c} \begin{array}{ccc} \Gamma & & \Gamma, [C] \\ \vdots & \equiv & \vdots \\ A & & D \end{array} \\ \hline C \to D \end{array}$$

by definition, if C is not evaluated in $\mathrm{Inf}^*(\mathcal{I})$, then $C \to D$ is evaluated in $\mathrm{Inf}^*(\mathcal{I})$; if C is evaluated in $\mathrm{Inf}^*(\mathcal{I})$, then, by induction hypothesis, D is evaluated in $\mathrm{Inf}^*(\mathcal{I})$. So, in general, $C \to D$ is evaluated in $\mathrm{Inf}^*(\mathcal{I})$.

- $\forall$ Introduction:

$$\begin{array}{c} \begin{array}{ccc} \Gamma & & \Gamma \\ \vdots & \equiv & \vdots\, (p) \\ A & & C(p) \end{array} \\ \hline \forall\, x.\, C(x) \ . \end{array}$$

If $C(t) \notin \mathrm{Inf}^*(\mathcal{I})$, for any term t, then $\forall\, x.\, C(x)$ is evaluated in $\mathrm{Inf}^*(\mathcal{I})$.

Let us suppose that there is a term t such that $C(t) \in \mathrm{Inf}^*(\mathcal{I})$; $\forall\, x.\, C(x)$ is evaluated in $\mathrm{Inf}^*(\mathcal{I})$ if and only if, $C(t)$ is evaluated in $\mathrm{Inf}^*(\mathcal{I})$.

Since $C(t) \in \mathrm{Inf}^*(\mathcal{I})$, there is an index k such that $C(t) \in \mathrm{Inf}(\mathrm{Exp}^k(\mathcal{I}))$, $\Gamma \subseteq \mathrm{Inf}(\mathrm{Exp}^k(\mathcal{I}))$ and $\begin{array}{c} \Gamma \\ \vdots \\ A \end{array} \prec \mathrm{Exp}^k(\mathcal{I})$, so,

by definition, $\begin{array}{c} \Gamma \\ \vdots\, (p:=t) \\ C(p) \end{array} \prec \mathrm{Exp}^{k+1}(\mathcal{I})$, that is, $\begin{array}{c} \Gamma \\ \vdots\, (p:=t) \\ C(p) \end{array} \prec$

$\mathrm{Exp}^*(\mathcal{I})$, and, by induction hypothesis, $C(t)$ is evaluated in $\mathrm{Inf}^*(\mathcal{I})$; from this fact, it follows that $\forall\, x.\, C(x)$ is evaluated in $\mathrm{Inf}^*(\mathcal{I})$.

- ∃ Elimination:

$$
\genfrac{}{}{0pt}{}{\begin{matrix}\Gamma\\ \vdots\\ A\end{matrix}}{} \equiv \frac{\begin{matrix}\Gamma\\ \vdots\\ \exists\, x.\, C(x)\end{matrix} \qquad \begin{matrix}\Gamma,[C(p)]\\ \vdots\, (p)\\ A\end{matrix}}{A}
$$

by induction hypothesis we know that $\exists\, x.\, C(x)$ is evaluated in $\mathrm{Inf}^*(\mathcal{I})$, and, by definition, there is a t such that $C(t)$ is evaluated in the same set, hence $C(t) \in \mathrm{Inf}^*(\mathcal{I})$.

Then, there is a k such that $\Gamma \cup \{C(t)\} \in \mathrm{Inf}(\mathrm{Exp}^k(\mathcal{I}))$

and $\begin{matrix}\Gamma\\ \vdots\\ A\end{matrix} \prec \mathrm{Exp}^k(\mathcal{I})$; so $\begin{matrix}\Gamma, C(p)\\ \vdots\, (p:=t)\\ A\end{matrix} \prec \mathrm{Exp}^{k+1}(\mathcal{I})$.

Thus, $\begin{matrix}\Gamma, C(p)\\ \vdots\, (p:=t)\\ A\end{matrix} \prec \mathrm{Exp}^*(\mathcal{I})$, and, by induction hypothesis, A is evaluated in $\mathrm{Inf}^*(\mathcal{I})$. $\qquad\square$

The conclusion we get from the previous lemmas is that every formula in $\mathrm{Inf}^*(\mathcal{I})$ gets evaluated. When this happens, remembering our intuitive reading of evaluation, it means that every collected formula is explained by the information content of the set of proofs $\mathcal{I}$.

Theorem 5.1.1 *Let $\mathcal{I}$ be a set of proofs, and let $A \in \mathrm{Inf}^*(\mathcal{I})$, then A is evaluated in $\mathrm{Inf}^*(\mathcal{I})$.*

Proof: From $A \in \mathrm{Inf}^*(\mathcal{I})$, we deduce that there is an index j such that $A \in \mathrm{Inf}(\mathrm{Exp}^j(\mathcal{I}))$.
Let's define a function $\mathrm{dg}(\cdot)$:

- $\mathrm{dg}_{\mathrm{Exp}^j(\mathcal{I})}(B) = 0$ if exists $\begin{matrix}\vdots\\ B\end{matrix} \prec \mathrm{Exp}^j(\mathcal{I})$ without undischarged assumptions;

- $\mathrm{dg}_{\mathrm{Exp}^j(\mathcal{I})}(B) = \max\{\mathrm{dg}_{\mathrm{Exp}^j(\mathcal{I})}(C_1), \ldots, \mathrm{dg}_{\mathrm{Exp}^j(\mathcal{I})}(C_n)\} +$
1 if there is a proof $\begin{matrix}C_1,\ldots,C_n\\ \vdots\\ B\end{matrix} \prec \mathrm{Exp}^j(\mathcal{I})$ and $\{C_1,\ldots,C_n\} \subseteq \mathrm{Inf}(\mathrm{Exp}^j(\mathcal{I}))$.

By induction on $dg_{Exp^j(\mathcal{I})}(\cdot)$, we prove that, for all $B \in Inf(Exp^j(\mathcal{I}))$, B is evaluated in $Inf^*(\mathcal{I})$:

- if $dg_{Exp^j(\mathcal{I})}(B) = 0$, then $\begin{array}{c} \vdots \\ B \end{array} \prec Exp^j(\mathcal{I})$, so $\begin{array}{c} \vdots \\ B \end{array} \prec Exp^*(\mathcal{I})$, and, by Lemma 5.1.2, B is evaluated in $Inf^*(\mathcal{I})$.

- if $dg_{Exp^j(\mathcal{I})}(B) > 0$, then $\begin{array}{c} C_1, \ldots, C_n \\ \vdots \\ B \end{array} \prec Exp^j(\mathcal{I})$.

So $\begin{array}{c} C_1, \ldots, C_n \\ \vdots \\ B \end{array} \prec Exp^*(\mathcal{I})$, but, by induction hypothesis, $C_1, \ldots, C_n$ are evaluated in $Inf^*(\mathcal{I})$, and, by Lemma 5.1.2, B is evaluated in $Inf^*(\mathcal{I})$, too.

Because $A \in Inf(Exp^j(\mathcal{I}))$, A is evaluated in $Inf^*(\mathcal{I})$. $\square$

The previous theorem is the key to link the global view of the information content of a proof, i.e., pseudo-truth sets, with its local view, i.e., the notion of evaluation.

Theorem 5.1.2 *Let $\mathcal{I}$ be a set of proofs, then their information content $Inf^*(\mathcal{I})$ is a pseudo-truth set.*

Proof: From the construction of $Inf^*(\mathcal{I})$,

- $A \in Inf^*(\mathcal{I})$ implies $\vdash A$, since it is the conclusion of a proof without undischarged assumption.

  In fact, by an immediate induction on the structure of the Coll operator, we get that there is proof with no undischarged assumptions that is composed by combining proofs in $Coll^*(\mathcal{I})$.

- $\neg B \in Inf^*(\mathcal{I})$ implies that $\vdash \neg B$, hence $\not\vdash B$, that implies $B \notin Inf^*(\mathcal{I})$, by construction.

- $B \vee C \in Inf^*(\mathcal{I})$ implies, by Theorem 5.1.1 that $B \vee C$ is evaluated in $Inf^*(\mathcal{I})$, and, by definition, B is evaluated in $Inf^*(\mathcal{I})$, or C is evaluated in $Inf^*(\mathcal{I})$, so $B \in Inf^*(\mathcal{I})$ or $C \in Inf^*(\mathcal{I})$.

- $B \wedge C \in \text{Inf}^*(\mathcal{I})$ implies, from Theorem 5.1.1 that $B \wedge C$ is evaluated in $\text{Inf}^*(\mathcal{I})$, and, by definition, B and C are evaluated in $\text{Inf}^*(\mathcal{I})$, so $B \in \text{Inf}^*(\mathcal{I})$ and $C \in \text{Inf}^*(\mathcal{I})$.

- $B \rightarrow C \in \text{Inf}^*(\mathcal{I})$ and $B \in \text{Inf}^*(\mathcal{I})$ implies that $B \rightarrow C$ and B are evaluated in $\text{Inf}^*(\mathcal{I})$, hence, by definition, C is evaluated in $\text{Inf}^*(\mathcal{I})$, so $C \in \text{Inf}^*(\mathcal{I})$.

- $\exists x . B(x) \in \text{Inf}^*(\mathcal{I})$ implies that $\exists x . B(x)$ is evaluated in $\text{Inf}^*(\mathcal{I})$, and, by definition, there is a term t such that $B(t)$ is evaluated in $\text{Inf}^*(\mathcal{I})$, so $B(t) \in \text{Inf}^*(\mathcal{I})$.

So, by definition, $\text{Inf}^*(\mathcal{I})$ is a pseudo-truth set. □

If we consider two special cases of the previous theorem, namely $\left\{ \begin{array}{c} \vdots \\ A \vee B \end{array} \right\}$ and $\left\{ \begin{array}{c} \vdots \\ \exists x . C(x) \end{array} \right\}$, we can immediately deduce that, for every disjunction, one of the disjunct appears in the information content of any proof for $A \vee B$, and, similarly, for any existential formula $\exists x . C(x)$, a witness for its truth can be found in the information content of its proof.

Hence, we our construction, embodied in the collection method, and its properties, allowed us to prove that **IL** enjoys the disjunction property and the explicit definability property, thus **IL** is naïvely constructive.

In addition, the collection method provides an effective operator, Inf, that extract the witnesses required to show the constructive character of the system. So, **IL** is more than just a constructive system, since there is an effective procedure, the collection method, which allows to construct the witnesses that are needed to explicit the constructive content of its proofs.

5.2   The E logic

The key idea underlying the collection method we tried to suggest in the previous exposition, is that the information extraction procedure is largely independent from the specific format of axioms and rules; the relevant notion is that the information we need is contained in the proofs.

In fact, the collection method for the **E** logic in its natural deduction presentation, is, essentially, the same as for **IL**. The definition of the Subproof operator is the obvious expansion of the definition we gave for **IL**; the Coll and Inf operators are the same as for **IL**.

The definition of the expansion operator Exp has to take in account also the negated parametric rules ¬∀E and ¬∃I.

Definition 5.2.1 *Let $\mathcal{I}$ be a set of proofs;*

$$\neg\forall E\text{-}Sub(\mathcal{I}) = \left\{ \left(\begin{array}{c} \Gamma, \neg A(p) \\ \vdots \\ B \end{array} \right) (p := t) \left| \begin{array}{cc} \Gamma & \Gamma, [\neg A(p)] \\ \vdots & \vdots \\ \neg\forall x. A(x) & B \\ \hline B \end{array} \right. \in \text{Coll}(\mathcal{I}) \wedge \Gamma \cup \{\neg A(t)\} \subseteq \text{Inf}(\mathcal{I}) \right\}$$

$$\neg\exists I\text{-}Sub(\mathcal{I}) = \left\{ \left(\begin{array}{c} \Gamma \\ \vdots \\ \neg A(p) \end{array} \right) (p := t) \left| \begin{array}{c} \Gamma \\ \vdots \\ \neg A(p) \\ \hline \neg\exists x. A(x) \end{array} \right. \in \text{Coll}(\mathcal{I}) \wedge \right.$$
$$\left. \wedge \Gamma \cup \{\neg A(t)\} \subseteq \text{Inf}(\mathcal{I}) \right\} \ .$$

Consequently

$$\text{Exp}(\mathcal{I}) = \mathcal{I} \cup \forall I\text{-}Sub(\mathcal{I}) \cup \exists E\text{-}Sub(\mathcal{I}) \cup$$
$$\cup \neg\forall E\text{-}Sub(\mathcal{I}) \cup \neg\exists I\text{-}Sub(\mathcal{I}) \ .$$

In the standard way one defines Exp*, Coll* and Inf*.

We want to observe here some important facts about the collection method:

- the exact definition of the Subproof operator depends on the calculus, but the notion of subproof is largely independent; as a result, for any calculus, the definition of the Subproof operator appears to be trivial.

- The definitions of the Coll and Inf operators are independent from the calculus; they depend just on the notion of subproof.

- The Exp operator depends on the substitution of eigen-variables in the proofs; we notice that the way Exp and the substitution operators ∀I-Sub, ∃E-Sub, ¬∃I-Sub and ¬∀E-Sub are defined does not really depend on the system, but they are instances of a common pattern.

- The □ rules, being non parametric, do not affect the construction of Coll* and Inf*.

The E logic is uniformly constructive

The proof stating that **E** is uniformly constructive follows the same pattern we have seen in Section 5.1 for **IL** . Hence we will just illustrate the additional notions, the differences and the reasons for them.

The notion of evaluation has to be enlarged because we have a new connective, □, and negation has a local semantics.

Definition 5.2.2 *A formula A is evaluated in a set of formulae $\mathcal{F}$, iff $A \in \mathcal{F}$ and*

- *A is a literal, i.e., atomic or negated atomic;*

- *$A \equiv \Box B$ or $A \equiv \neg\Box B$;*

- *$A \equiv B \wedge C$ and B and C are both evaluated in $\mathcal{F}$;*

- *$A \equiv \neg(B \wedge C)$ and one of $\neg B$, $\neg C$ is evaluated in $\mathcal{F}$;*

- *$A \equiv B \vee C$ and B is evaluated in $\mathcal{F}$ or C is evaluated in $\mathcal{F}$;*

- *$A \equiv \neg(B \vee C)$ and $\neg B$ and $\neg C$ are both evaluated in $\mathcal{F}$;*

- *$A \equiv B \rightarrow C$ and if B is evaluated in $\mathcal{F}$ then also C is;*

- *$A \equiv \neg(B \rightarrow C)$ and B and $\neg C$ are both evaluated in $\mathcal{F}$;*

- *$A \equiv \neg\neg B$ and B is evaluated in $\mathcal{F}$;*

- *$A \equiv \forall x. B(x)$ and, for every term t such that $B(t) \in \mathcal{F}$, $B(t)$ is evaluated in $\mathcal{F}$;*

- *$A \equiv \neg\forall x. B(x)$ and there is a term t such that $\neg B(t)$ is evaluated in $\mathcal{F}$;*

- $A \equiv \exists\, x.\, B(x)$ and there is a term t such that $B(t)$ is evaluated in $\mathcal{F}$;

- $A \equiv \neg\exists\, x.\, B(x)$ and, for every term t such that $\neg B(t) \in \mathcal{F}$, $\neg B(t)$ is evaluated in $\mathcal{F}$.

From this definition it is easy to prove

Lemma 5.2.1 *Let $\mathcal{I}$ be a set of proofs, if* $\begin{smallmatrix}\Gamma\\[-2pt]\vdots\\[-2pt]A\end{smallmatrix}$ *$\prec \mathrm{Exp}^*(\mathcal{I})$ and $\Gamma \subseteq$*

$\mathrm{Inf}^*(\mathcal{I})$, *then $A \in \mathrm{Inf}^*(\mathcal{I})$.*

Proof: See Lemma 5.1.1. □

Lemma 5.2.2 *Let $\mathcal{I}$ be a set of proofs, let* $\begin{smallmatrix}\Gamma\\[-2pt]\vdots\\[-2pt]A\end{smallmatrix}$ *$\prec \mathrm{Exp}^*(\mathcal{I})$ and let,*

for every $\gamma \in \Gamma$, γ be evaluated in $\mathrm{Inf}^*(\mathcal{I})$, then A is evaluated in $\mathrm{Inf}^*(\mathcal{I})$.

Proof: See Lemma 5.1.2. For the remaining cases in the induction,

- $\neg\wedge$, $\neg\vee$, $\neg\rightarrow$, $\neg\neg$ and $\neg\forall$ Introduction, $\neg\vee$, $\neg\rightarrow$, $\neg\neg$ and $\neg\exists$ Elimination: by induction hypothesis, the conclusions of the immediate subproofs are evaluated, thus, by definition, the conclusion is evaluated, too.

- $\neg\wedge$ Elimination:

$$
\begin{smallmatrix}\Gamma\\[-2pt]\vdots\\[-2pt]A\end{smallmatrix}
\equiv
\dfrac{\begin{smallmatrix}\Gamma\\[-2pt]\vdots\\[-2pt]\neg(B\wedge C)\end{smallmatrix} \quad \begin{smallmatrix}\Gamma,[\neg B]\\[-2pt]\vdots\\[-2pt]A\end{smallmatrix} \quad \begin{smallmatrix}\Gamma,[\neg C]\\[-2pt]\vdots\\[-2pt]A\end{smallmatrix}}{A}
$$

By induction hypothesis, $\neg(B \wedge C)$ is evaluated, then $\neg B$ is evaluated or $\neg C$ is evaluated. In the former case, applying the induction hypothesis to $\begin{smallmatrix}\Gamma,\neg B\\[-2pt]\vdots\\[-2pt]A\end{smallmatrix}$, we get that A is evaluated; in the latter case, applying the induction hypothesis to $\begin{smallmatrix}\Gamma,\neg C\\[-2pt]\vdots\\[-2pt]A\end{smallmatrix}$, we get that A is evaluated.

- $\neg\forall$ Elimination:

$$\begin{array}{c}\Gamma\\\vdots\\A\end{array}\;\equiv\;\dfrac{\begin{array}{cc}\begin{array}{c}\Gamma\\\vdots\\\neg\forall\,x.\,B(x)\end{array}&\begin{array}{c}\Gamma,[\neg B(p)]\\\vdots\\A\end{array}\end{array}}{A}$$

By induction hypothesis $\neg\forall\,x.\,B(x)$ is evaluated, so there is an index k and a term t such that $\begin{array}{c}\Gamma\\\vdots\\A\end{array}\prec\mathrm{Exp}^k(\mathcal{I})$ and $\Gamma\cup\{B(t)\}\subseteq\mathrm{Inf}(\mathrm{Exp}^k(\mathcal{I}))$.

Then $\left(\begin{array}{c}\Gamma,\neg B(p)\\\vdots\\A\end{array}\right)(p:=t)\in\mathrm{Exp}^{k+1}(\mathcal{I})$.

Thus $\left(\begin{array}{c}\Gamma,\neg B(p)\\\vdots\\A\end{array}\right)(p:=t)\in\mathrm{Exp}^*(\mathcal{I})$.

Applying the induction hypothesis, we get that A is evaluated in $\mathrm{Inf}^*(\mathcal{I})$.

- $\neg\exists$ Introduction:

$$\begin{array}{c}\Gamma\\\vdots\\A\end{array}\;\equiv\;\dfrac{\begin{array}{c}\Gamma\\\vdots\\\neg B(p)\end{array}}{\neg\exists\,x.\,B(x)}$$

For every term t such that $\neg B(t)\in\mathrm{Inf}^*(\mathcal{I})$, there is an index k such that $\begin{array}{c}\Gamma\\\vdots\\A\end{array}\prec\mathrm{Exp}^k(\mathcal{I})$ and $\Gamma\cup\{\neg B(t)\}\subseteq\mathrm{Inf}(\mathrm{Exp}^k(\mathcal{I}))$, hence $\left(\begin{array}{c}\Gamma\\\vdots\\\neg B(p)\end{array}\right)(p:=t)\in\mathrm{Exp}^{k+1}(\mathcal{I})$.

Thus, $\left(\begin{array}{c}\Gamma\\\vdots\\\neg B(p)\end{array}\right)(p:=t)\in\mathrm{Exp}^*(\mathcal{I})$.

By induction hypothesis it follows that $\neg B(t)$ is evaluated. So, $\neg\exists\,x.\,B(x)$ is evaluated in $\mathrm{Inf}^*(\mathcal{I})$. $\neg B(t)$ is evaluated in $\mathrm{Inf}^*(\mathcal{I})$, by definition.

- $\neg$ Elimination:

$$
\begin{array}{c}
\Gamma \\ \vdots \\ A
\end{array}
\equiv
\begin{array}{cc}
\begin{array}{c}\Gamma \\ \vdots \\ B\end{array} & \begin{array}{c}\Gamma \\ \vdots \\ \neg B\end{array} \\ \hline & P
\end{array}
\quad \text{or} \quad
\begin{array}{c}
\Gamma \\ \vdots \\ A
\end{array}
\equiv
\begin{array}{cc}
\begin{array}{c}\Gamma \\ \vdots \\ B\end{array} & \begin{array}{c}\Gamma \\ \vdots \\ \neg B\end{array} \\ \hline & \neg P
\end{array}
$$

Being P atomic, P and $\neg P$ are evaluated by definition.

- $\square$ Introduction, $\square$ Elimination:

$$
\begin{array}{c}
\Gamma \\ \vdots \\ A
\end{array}
\equiv
\begin{array}{cc}
\begin{array}{c}\Gamma,[\neg B] \\ \vdots \\ C\end{array} & \begin{array}{c}\Gamma,[\neg B] \\ \vdots \\ \neg C\end{array} \\ \hline & \square B
\end{array}
\quad \text{or} \quad
\begin{array}{c}
\Gamma \\ \vdots \\ A
\end{array}
\equiv
\begin{array}{cc}
\begin{array}{c}\Gamma,[B] \\ \vdots \\ C\end{array} & \begin{array}{c}\Gamma,[B] \\ \vdots \\ \neg C\end{array} \\ \hline & \neg \square B
\end{array}
$$

Since the conclusion is a boxed formula, it is evaluated in $\mathrm{Inf}^*(\mathcal{I})$. $\qquad\square$

Theorem 5.2.1 *Let $\mathcal{I}$ be a set of proofs, and let $A \in \mathrm{Inf}^*(\mathcal{I})$, then A is evaluated in $\mathrm{Inf}^*(\mathcal{I})$.*

Proof: See Theorem 5.1.1. $\qquad\square$

At this point, we can prove, exactly in the same way as for **IL**, Theorem 5.1.2, and so **E** is uniformly constructive. But we prefer to modify the definition of pseudo-truth sets to give a proof which provides a stronger theorem, taking into account the validity of De Morgan's laws.

Definition 5.2.3 *A set $\mathcal{F}$ of formulae is a* pseudo-truth set with negation *iff*

- $A \in \mathcal{F}$ implies $\vdash A$;

- $A \vee B \in \mathcal{F}$ implies $A \in \mathcal{F}$ or $B \in \mathcal{F}$;

- $\neg(A \vee B) \in \mathcal{F}$ implies $\neg A \in \mathcal{F}$ and $\neg B \in \mathcal{F}$;

- $A \wedge B \in \mathcal{F}$ implies $A \in \mathcal{F}$ and $B \in \mathcal{F}$;

- $\neg(A \wedge B) \in \mathcal{F}$ implies $\neg A \in \mathcal{F}$ or $\neg B \in \mathcal{F}$;

- $A \rightarrow B \in \mathcal{F}$ implies if $A \in \mathcal{F}$, then $B \in \mathcal{F}$;

- $\neg(A \rightarrow B) \in \mathcal{F}$ implies $A \in \mathcal{F}$ and $\neg B \in \mathcal{F}$;

- $\exists x. A(x) \in \mathcal{F}$ implies that there is a t such that $A(t) \in \mathcal{F}$;

- $\neg \forall x. A(x) \in \mathcal{F}$ implies that there is a term t such that $\neg A(t) \in \mathcal{F}$.

Theorem 5.2.2 *Let $\mathcal{I}$ be a set of proofs, then* $\mathrm{Inf}^*(\mathcal{I})$ *is a pseudo-truth set with negation.*

Proof: See Theorem 5.1.2. The extra cases are immediate consequences of the notion of evaluation. □

5.3 □-theories

The notion of □-theory has been introduced in Chapter 3 to model the conservative expansions of specification frameworks. The following facts can be proved in **E**:

- $\Box(A \land B) \leftrightarrow \Box A \land \Box B$;

- $\Box(A \rightarrow B) \leftrightarrow (A \rightarrow \Box B)$;

- $\Box \forall x. A \leftrightarrow \forall x. \Box A$.

Hence, by induction on the structure of E-Harrop formulae, one proves

Lemma 5.3.1 *For any E-Harrop formula H, $\vdash \Box H \leftrightarrow H$.*

One may notice that the fact $\Box \forall x. A \leftrightarrow \forall x. \Box A$ is an alternative way to express the Kuroda principle in the **E** logic; in fact the □ connective acts similarly to double negation in **IL**. We remind that Kuroda logic is the extension of the **IL** system plus the aforementioned Kuroda principle.

The importance of □-theories comes from the way we formalise datatypes, since we imposed that the axioms of an open framework are E-Harrop formulae; hence, we want an instance of the collection method permitting to extract information from a system $\mathbf{E} + T$, where T is a □-theory.

We consider the system $\mathbf{E} + T$ composed by the same inference rules as **E** plus

$$\frac{-\phi \in T}{\phi}.$$

The definitions of the basic operators, Subproof, Coll and Inf are the same as for the **E** logic. The definition of the Exp operator is enlarged to take into account the peculiar nature of the □-theory T; we define

Definition 5.3.1

$$Har_\wedge\text{-}Sub(\mathcal{I}) = \left\{ \frac{A \wedge B}{A}, \frac{A \wedge B}{B} \,\middle|\, A \wedge B \in T \right\}$$

$$Har_\to\text{-}Sub(\mathcal{I}) = \left\{ \frac{A \quad A \to B}{B} \,\middle|\, A \to B \in T \wedge A \in \mathrm{Inf}(\mathcal{I}) \right\}$$

$$Har_\forall\text{-}Sub(\mathcal{I}) = \left\{ \frac{\forall x. A(x)}{A(t)} \,\middle|\, \forall x. A(x) \in T \wedge \right.$$

$$\left. \wedge A(t) \in \mathrm{Inf}(\mathcal{I}) \right\}$$

$$Har\text{-}Sub(\mathcal{I}) = \quad Har_\wedge\text{-}Sub(\mathcal{I}) \cup Har_\to\text{-}Sub(\mathcal{I}) \cup$$
$$\cup Har_\forall\text{-}Sub(\mathcal{I}) .$$

Thus, calling $\mathrm{Exp}_\mathbf{E}$ the expansion operator for the **E** logic, the Exp operator becomes

$$\mathrm{Exp}(\mathcal{I}) = \mathrm{Exp}_\mathbf{E}(\mathcal{I}) \cup \mathrm{Har\text{-}Sub}(\mathcal{I}) .$$

Then, in the standard way, we define Exp^*, Coll^* and Inf^*.

The idea behind the definition of the Har-Sub operator is the same as we have seen for Exp_IL: the expansion operator provides a way to *dismount* the subproofs of a set of proofs $\mathcal{I}$ which cannot be managed by the Coll operator. One way is to instantiate eigenvariables, another way is to apply the appropriate elimination rules to the E-Harrop axioms.

□-theories are uniformly constructive

In this section we prove that extending **E** with a set $\mathcal{H}$ of E-Harrop axioms, we get a uniformly constructive formal system. As for other theories and logics we have presented in this chapter, the only part of the proof leading to state that $\mathbf{E} + \mathcal{H}$ is uniformly constructive that has to be changed is the induction in Lemma 5.1.2.

We just report the new case:

- **E-Harrop Axiom:**

$$\begin{array}{c} \Gamma \\ \vdots \\ A \end{array} \equiv \frac{}{A} \, , \, A \in \mathcal{H}$$

By induction on the structure of the formula A, we prove that A is evaluated in $\mathrm{Inf}^*(\mathcal{I})$:

- if A is atomic then it is evaluated in $\mathrm{Inf}^*(\mathcal{I})$.

- if $A \equiv \neg B$ and B is atomic then, by definition, it is evaluated in $\mathrm{Inf}^*(\mathcal{I})$.

- if $A \equiv \square B$ then, it is evaluated in $\mathrm{Inf}^*(\mathcal{I})$.

- if $A \equiv B \wedge C$ then, by definition, B and C are in $\mathrm{Inf}^*(\mathcal{I})$, thus, by induction hypothesis, being E-Harrop formulae, both are evaluated, and so also $B \wedge C$ is evaluated in $\mathrm{Inf}^*(\mathcal{I})$.

- if $A \equiv B \rightarrow C$, if B is evaluated in $\mathrm{Inf}^*(\mathcal{I})$, then $B \in \mathrm{Inf}^*(\mathcal{I})$, hence, by definition of Har-Sub, $C \in \mathrm{Inf}^*(\mathcal{I})$. But, by induction hypothesis, being an E-Harrop, formula, C is evaluated in $\mathrm{Inf}^*(\mathcal{I})$, then $B \rightarrow C$ is evaluated in $\mathrm{Inf}^*(\mathcal{I})$, too.

- if $A \equiv \forall x . B(x)$, then, for every term t such that $B(t) \in \mathrm{Inf}^*(\mathcal{I})$, since $B(t)$ is an E-Harrop formula, by induction hypothesis, $B(t)$ is evaluated. Hence, by definition, $\forall x . B(x)$ is evaluated in $\mathrm{Inf}^*(\mathcal{I})$. $\square$

By looking at the proofs, one should notice that a similar result holds for Harrop theories with respect to **IL**.

5.4 Identity

The axioms about equality have a special role in logic, because the intended meaning of this relation is fixed. The theory **ID** for equality is composed by an axiom and a rule:

$$\frac{}{t = t} \, \mathrm{Refl} \qquad \frac{x = y \quad A(y)}{A(x)} \, \mathrm{Sub}$$

Usually, and this is the case for **IL** and **E**, we can impose that $A(x)$ in the Sub rule is atomic; then we can prove the general case by induction on the structure of formulae.

To extract information from the system $L + $ **ID**, where L is any logical theory to which the collection method can be applied to prove it is uniformly constructive, we build, in the standard way, the Subproof, the Coll and the Inf operators.

As usual, we have to enlarge the Exp_L operator:

Definition 5.4.1 *Let $\mathcal{I}$ be a set of proofs, the substitution operator for the* **ID** *theory is defined as*

$$\text{ID-Sub}(\mathcal{I}) = \left\{ \frac{t = s \quad A(s)}{A(t)} \;\middle|\; \{t = s, A(s)\} \subseteq \text{Inf}(\mathcal{I}) \right\} .$$

So, $\text{Exp}(\mathcal{I})$ becomes $\text{Exp}_L(\mathcal{I}) \cup \text{ID-Sub}(\mathcal{I})$.

Then, we construct $\text{Exp}^*(\mathcal{I})$, $\text{Coll}^*(\mathcal{I})$ and $\text{Inf}^*(\mathcal{I})$ along the now usual guidelines.

Identity theory is uniformly constructive

We assume that the proof that L is uniformly constructive follows the schema we employed in all other cases.

The proof that $L + $ **ID** is uniformly constructive, is identical to the one for L, except, as usual, two new cases in the induction of Lemma 5.1.2:

- Reflexivity:

$$\begin{array}{c} \Gamma \\ \vdots \\ A \end{array} \equiv \overline{\; t = t \;}$$

Since $t = t$ is atomic, it is evaluated in $\text{Inf}^*(\mathcal{I})$.

- Substitution:

$$\begin{array}{c} \Gamma \\ \vdots \\ A \end{array} \equiv \frac{\begin{array}{cc} \Gamma & \Gamma \\ \vdots & \vdots \\ t = s & a(s) \end{array}}{a(t)}$$

Again, by definition of the substitution rule, $a(t)$ is an atomic formula, and, thus, it gets evaluated in $\text{Inf}^*(\mathcal{I})$. $\square$

Before, we introduced the ID-Sub operator; it is not necessary when we adopt the restricted version (with atomic conclusion) of the substitution rule, but it is the device permitting

$$\frac{}{\neg s(x) = 0} \quad \frac{}{s(x) = s(y) \to x = y}$$

$$\frac{}{x + 0 = x} \quad \frac{}{x + s(y) = s(x + y)}$$

$$\frac{}{x \cdot 0 = 0} \quad \frac{}{x \cdot s(y) = x + x \cdot y}$$

$$\begin{array}{c} [A(p)] \\ \vdots \\ \frac{A(0) \quad A(s(p))}{A(t)} \text{ Ind}(*) \end{array}$$

where, in $(*)$ p is an eigenvariable.

Table 5.1: Inference rules of Peano arithmetic.

to perform the proof if we have only the unrestricted substitution rule. We leave the proof of that case to the reader, since it is just an induction on the structure of formulae.

5.5 Induction principles

As we have seen in Chapter 3, most theories need one or more induction principles; in Chapter 4, we gave them a computational reading by showing how they become program schemata involving recursion or cycles.

To treat induction principles in our method we should keep in mind that induction rules are *parametric*, i.e., they have eigenvariables.

As we have already seen, we treat eigenvariables by enlarging the definition of the Exp operator via suitable substitution rules that instantiate the eigenvariables, eventually adding proofs that make explicit some aspects of the inductive construction which is formalised in the rule.

In the following we will show how to treat induction in Peano arithmetic, and the descending chain principle. These two examples will clarify the general technique, and, at the same time, they show that our infrastructure, as developed in the previous lectures, is homogeneous.

Peano arithmetic is formalised in the usual way, as shown in Table 5.1. We notice that the logical system $\mathbf{E} + \mathbf{ID} + \mathbf{PA}$

can be reduced to $\mathbf{E} + \mathbf{ID} + T + \text{Ind}$, where T is a $\Box$-theory and Ind is the induction principle on natural numbers.

The way to treat induction is to enlarge the Exp operator as defined for the system $\mathbf{E} + \mathbf{ID} + T$; let us call it $\text{Exp}_{\mathbf{E}+\mathbf{ID}+T}$.

Definition 5.5.1 *Let $\mathcal{I}$ be a set of proofs, we define Ind-Sub$(\mathcal{I})$ as the smallest set such that, if,*

$$
\cfrac{
\begin{array}{ccc}
\Gamma & & \Gamma, [A(p)] \\
\vdots & & \vdots \ (p) \\
A(0) & & A(s(p))
\end{array}
}{A(t)} \ \text{Ind} \quad \prec \mathcal{I}
$$

with t a closed term, and $\Gamma \cup \{A(t)\} \subseteq \text{Inf}(\mathcal{I})$, then

$$
\left\{
\begin{array}{c}
\vdots \\
t = s^i(0)'
\end{array}
\quad
\begin{array}{c}
t = s^i(0), A(s^i(0)) \\
\vdots \\
A(t)
\end{array}
\right\} \cup
$$

$$
\cup \left\{ \left(\begin{array}{c} \Gamma, A(p) \\ \vdots \\ A(s(p)) \end{array} \right) (p := s^j(0)) \ \middle| \ 0 \le j < i \right\} \subseteq \text{Ind-Sub}(\mathcal{I}) \ ,
$$

where $\begin{array}{c} \vdots \\ t = s^i(0) \end{array}$ *is any proof that converts t to its canonical form, i.e., as a numeral.*

We remark that there is a standard way to perform the

proof $\begin{array}{c} \vdots \\ t = s^i(0) \end{array}$ (the reader is invited to try to prove this fact).

In this way, in the system $\mathbf{E} + T + \text{Ind}$, one defines

$$
\text{Exp}(\mathcal{I}) = \text{Exp}_{\mathbf{E}+T}(\mathcal{I}) \cup \text{Ind-Sub}(\mathcal{I}) \ .
$$

Then, as usual, we can construct $\text{Exp}^*(\mathcal{I})$, $\text{Coll}^*(\mathcal{I})$ and $\text{Inf}^*(\mathcal{I})$, for any set of proofs $\mathcal{I}$.

As we already noticed, Peano arithmetic is a $\Box$-theory plus the induction rule, and we know that adding to a uniformly constructive system, like $\mathbf{E} + \mathbf{ID}$, a $\Box$-theory leads us to a uniformly constructive system; now we will prove that adding also the induction principle does not modify this character of the theory

On closed proofs, it is possible to prove that $\mathbf{E} + \mathbf{ID} + \mathbf{PA}$ is uniformly constructive: the reason for this fact lies in the definition of the Ind-Sub operator, where we need to fix a numeral which is the *value* of t, the term up to which we use the induction. Technically, the proof that $\mathbf{E} + \mathbf{ID} + \mathbf{PA}$ is uniformly constructive is the same as for $\mathbf{E} + \mathbf{ID}$ plus a $\Box$-theory, except for Lemma 5.1.2 where, having another inference rule, we need a new case:

- Induction:

$$\begin{array}{c} \Gamma \\ \vdots \\ A \end{array} \equiv \cfrac{\begin{array}{ccc} & \Gamma & \Gamma, [B(p)] \\ & \vdots & \vdots \; {\scriptstyle (p)} \\ B(0) & & B(\mathbf{s}(p)) \end{array}}{B(t)}$$

Let i be a number such that $\mathbf{s}^i(0) = t$ is provable; by induction on k, $0 \leq k \leq i$, we prove that $B(\mathbf{s}^k(0))$ gets evaluated in $\mathrm{Inf}^*(\mathcal{I})$:

  - $k = 0$: $B(0)$ is evaluated by the primary induction hypothesis.
  - $K = \mathbf{s}(k')$: by the secondary induction hypothesis $B(k')$ is evaluated in $\mathrm{Inf}^*(\mathcal{I})$; applying the primary induction hypothesis to the induction step, and remembering the definition of Ind-Sub, one gets that $B(\mathbf{s}(k'))$ is evaluated in $\mathrm{Inf}^*(\mathcal{I})$.

So $B(\mathbf{s}^i(0))$ is evaluated in $\mathrm{Inf}^*(\mathcal{I})$, but, by definition of Ind-Sub, $\mathbf{s}^i(0) = t \in \mathrm{Inf}^*(\mathcal{I})$, and, being atomic, it is also evaluated. From the definition of Ind-Sub again,

$$\cfrac{\mathbf{s}^i(0) = t \quad B(\mathbf{s}^i(0))}{B(t)} \in \mathrm{Coll}^*(\mathcal{I}) \; ,$$

thus, by induction hypothesis, $B(t)$ gets evaluated in $\mathrm{Inf}^*(\mathcal{I})$. $\qquad\Box$

Hence, $\mathbf{E} + \mathbf{ID} + \mathbf{PA}$ is uniformly constructive on closed formulae; of course, one can prove on the same guidelines the $\mathbf{IL} + \mathbf{ID} + \mathbf{PA}$ is uniformly constructive on closed formulae.

The descending chain principle is treated in a similar way; let us suppose to work in the logical system $\mathbf{E} + T + \mathrm{DCP}$, where T is any $\square$-theory whose signature contains a binary relational symbol $<$, and let us suppose that the intended model of T makes $<$ to be interpreted as a well-founded ordering.

The descending chain principle has the shape:

$$
\begin{array}{c}
[B(p)] \\
\vdots \\
\dfrac{\exists\, x.\, B(x) \quad (\exists\, z.\, B(z) \wedge z < p) \vee A}{A} \quad \mathrm{DCP}
\end{array}
$$

As before, being DCP a parametric rule, we expand the definition of $\mathrm{Exp}_{\mathbf{E}+T}$:

Definition 5.5.2 *Let $\mathcal{I}$ be a set of closed proofs*

$$
DCP\text{-}Sub(\mathcal{I}) = \left\{ \left(\begin{array}{c} \Gamma, B(p) \\ \vdots \\ (\exists\, z.\, B(z) \wedge z < p) \vee A \end{array} \right) (p := t) \;\middle|\; \right.
$$

$$
\Gamma \cup \{B(t)\} \subseteq \mathrm{Inf}(\mathcal{I}) \wedge
$$

$$
\left. \wedge \dfrac{\begin{array}{cc} \Gamma & [B(p)] \\ \vdots & \vdots \\ \exists\, x.\, B(x) & (\exists\, z.\, B(z) \wedge z < p) \vee A \end{array}}{A} \in \mathrm{Coll}(\mathcal{I}) \right\}.
$$

The Exp operator becomes

$$
\mathrm{Exp}(\mathcal{I}) = \mathrm{Exp}_{\mathbf{E}+T}(\mathcal{I}) \cup DCP\text{-}Sub(\mathcal{I})\;,
$$

and, as usual, we construct Exp^*, Coll^* and Inf^*.

We want to prove that the system $L + \mathrm{DCP}$, where L is a uniformly constructive system in which the $<$ relation is a well-ordering, is uniformly constructive.

As the reader may suppose, the proof that L is uniformly constructive follows the usual schema; adding the DCP rule does not modify the proof, except for the Lemma 5.1.2, where a new case in the main induction has to be considered:

- Descending chain principle:

$$
\begin{array}{c}
\Gamma \\ \vdots \\ A
\end{array}
\equiv
\dfrac{
\begin{array}{ccc}
\Gamma & \Gamma & \Gamma,[B(p)] \\
\vdots & \vdots & \vdots \\
\exists\, x.\, B(x) & (\exists\, z.\, B(z) \wedge z < p) \vee A
\end{array}
}{A}
$$

By induction hypothesis, $\exists\, x.\, B(x)$ is evaluated, so there is a term t such that $B(t)$ is evaluated.

Hence, we can find an index k such that $\begin{array}{c}\Gamma \\ \vdots \\ A\end{array} \prec \mathrm{Exp}^k(\mathcal{I})$

and $\Gamma \cup \{B(t)\} \subseteq \mathrm{Inf}(\mathrm{Exp}^k(\mathcal{I}))$, thus

$$
\left(
\begin{array}{c}
\Gamma, B(p) \\ \vdots \\ (\exists\, z.\, B(z) \wedge z < p) \vee A
\end{array}
\right)
(p := t) \in \mathrm{Exp}^{k+1}(\mathcal{I}) \ .
$$

Applying the induction hypothesis again, we get that $(\exists\, z.\, B(z) \wedge z < p) \vee A$ is evaluated, i.e., $\exists\, z.\, B(z) \wedge z < p$ is evaluated or A is evaluated. In the latter case, we are done; in the former, we know that there is a term t' such that $B(t') \wedge t' < p$ is evaluated, i.e., $B(t')$ and $t' < t$ are both evaluated.

Iterating the reasoning on t', we have a sequence of terms, $t, t', \ldots$, for which $B(t), B(t'), \ldots$ are evaluated and $t' < t, t'' < t', \ldots$ are evaluated, hence they are true since they are proved.

But $<$ is a well ordering relation on the domain, so this sequence of terms cannot be infinite, and, for this reason, eventually A gets evaluated in $\mathrm{Inf}^*(\mathcal{I})$.

Then, in the same way as for L, it follows that $L + \mathrm{DCP}$ is uniformly constructive.

5.6 Further remarks

In the previous sections, we have proved that many logical systems are uniformly constructive.

The formal machinery we presented, called the collection method, and the general schema for proving a system to be uniformly constructive, is far more powerful: it is possible to prove that many other logics and theories are uniformly constructive, essentially in the same way as we did till now.

We invite the reader to try, e.g., with **IL** + Kur, that is, Kuroda logic [20, 2, 34, 35], where the Kur rule is

$$\frac{\forall x. \neg\neg A(x)}{\neg\neg\forall x. A(x)}$$

or with **IL** + $\mathcal{H}$, where $\mathcal{H}$ is a set of Harrop formulae.

The proving technique we adopted is not the only way to prove that a system is uniformly constructive; in [12] another technique is presented that leads to proofs for the Kreisel-Putnam logic, the Scott logic, the Grzegorczyck logic and the Markov arithmetic. All these systems are problematic with the collection method approach.

As a final remark, it is important to underline that the notion of uniformly constructive formal system has a proper logical meaning, we borrowed to understand the inner relation between a constructive and a computational view of specification formulae, a relation we assumed so far.

Chapter 6

Program Analysis

The goal of this chapter is to introduce the reader to the formal analysis of programs. Supposing to have a program and a correctness proof for it, the problem we would like to answer is: "what kind of information can be extracted from the correctness proof in order to gain a better understanding of the program behaviour?"

The ultimate goal is to bridge the gap between the knowledge used by the programmer and the knowledge developed in the proof. The former knowledge comes from the programming activity, usually not conducted in a formal setting and closer to a craft than to a method; the latter knowledge is based on a purely formal reasoning.

The distance between the two knowledges brings to an underestimation of the value of a correctness proof, since it is difficult to develop, expensive, and hard to understand. One of the goals of the formal analysis of programs is to extract from a correctness proof enough information to convince an human expert of its exactness. In fact, establishing a correctness proof to be flawless is a problem: it is easy, by means of a proof checker, to guarantee that the proof does not contain logical errors, but it is very difficult to certify that the formal specifications it is based on correspond to the "real" specifications, the ones describing the problem to be solved.

Our approach to program analysis is quite direct: we use the collection method to synthesise the information content of the given correctness proof, and, then, we try to automatically extract from the information content a family of interesting facts about the program.

6.1 Translating object code into logic

The first step we need to perform when trying to formally analyse a program, is to transform the program's code into a mathematical object we can treat.

In this section, we concentrate on object code programs, that is, programs written in the native machine language of some CPU. Our choice is to formalise the assembly language of the MC68000 microprocessor.

In fact, choosing an assembly language simplifies the formalisation process: every microprocessor has a data book that describes in every detail the meaning of each machine instruction by means of the transformation it induces on the memory and on the registers. Thus, it is easy to formalise the semantics of a machine language into an operational semantics whose state is given by the computer memory and the microprocessor registers. Since the assembly language is just the human-readable form of the machine language, it is immediate to formalise the symbolic form, when we know how to formalise the machine language.

The formalisation process is described by means of a mechanical translation procedure that takes a piece of object code as input and produces a logical representation for it. The translation procedure takes as input an assembly source code where no macros are present and where every address is resolved, and it translates this code in a logical representation.

The translation algorithm operates on the language of **IL** plus modular arithmetic where the basic scalar types byte, word and longword (integers modulo 2^8, modulo 2^{16} and modulo 2^{32}, respectively) have been defined in a suitable closed framework. The output of the translation procedure is a logical theory, the *program theory*, containing a series of axioms, one per instruction, encoding the program. The program theory is formally represented as an open specification framework that depends on the *microprocessor theory*, that acts as the parameter of the open framework.

The theory of the microprocessor has three roles:

- it provides the minimal set of instruments to reason about object code programs;

- it declares the types needed to represent the code;

- it declares the constants that constitute the world the microprocessor operates on.

The types are byte, word and longword, used to represent the quantities the microprocessor operates on, and time, that is used to model how the flow of control is passed from one instruction to another.

The types byte, word and longword are specialisations of modular numbers, and specifically, byte $=$ Int$/$ (mod 2^8), word $=$ Int$/$ (mod 2^{16}), longword $=$ Int$/$ (mod 2^{32}); we use both signed and unsigned bytes (words, longwords, respectively), thus our types are sbyte for signed bytes (values from -128 to 127), and ubyte for unsigned bytes (values from 0 to 255). An analogous notation is used for signed (unsigned) words and longwords.

The type time, following the fact that the microprocessor clock is discrete, is modelled by integer numbers.

In the microprocessor theory, the constants for memory and registers are declared. Specifically, the MC68000 microprocessor provides sixteen registers, eight of them being data registers, the others being address registers. Addresses in the memory are represented by means of unsigned longwords. The details of the MC68000 architecture can be found in [43].

We model registers as functions from time to values:

$$d_i : \text{time} \rightarrow \text{slongword} \quad ,0 \leq i \leq 7$$
$$a_i : \text{time} \rightarrow \text{slongword} \quad ,0 \leq i \leq 7$$
$$\text{pc}: \text{time} \rightarrow \text{ulongword}$$

A particular case is the status register that is modelled by a set of functions, one for each flag in the register:

$$\text{Zflag}: \text{time} \rightarrow \text{bool} \quad (* \text{ zero } *)$$
$$\text{Nflag}: \text{time} \rightarrow \text{bool} \quad (* \text{ negative } *)$$
$$\text{Cflag}: \text{time} \rightarrow \text{bool} \quad (* \text{ carry } *)$$
$$\text{Vflag}: \text{time} \rightarrow \text{bool} \quad (* \text{ overflow } *)$$
$$\text{Xflag}: \text{time} \rightarrow \text{bool} \quad (* \text{ extension } *)$$

The memory is represented as a function from addresses and times to values:

$$\texttt{memory}: \texttt{ulongword} \times \texttt{time} \rightarrow \texttt{byte}$$

It is quite handy to define predicates for reading and writing bytes, words and longwords in memory. We leave to the reader their elementary definitions [6].

Every instruction becomes a logical axiom and these axioms are grouped in the program theory. The general format of the logical representation of an instruction I is

$$\forall t: \texttt{time}.\, \texttt{pc}(t) = A \rightarrow B \wedge C$$

where A is the address of the instruction I, B specifies the value of the program counter at time $t + 1$, and C specifies the value of every register, flag and memory cell at time $t + 1$, depending on the instruction operands, the status of memory at time t, and the values of registers and flags at time t.

The format of the B part can be either

$$\texttt{pc}(t + 1) = H(\texttt{pc}(t))$$

or

$$(f(t) \rightarrow \texttt{pc}(t + 1) = H_1(\texttt{pc}(t))) \wedge$$
$$\wedge\, (\neg f(t) \rightarrow \texttt{pc}(t + 1) = H_2(\texttt{pc}(t)))$$

where H, H_1 and H_2 are expressions depending on the current value of the program counter and calculating the address of the next instruction to execute; $f(t)$ is a test formula, depending on the time t, generally, a conjunction/disjunction of (negations of) flag predicates.

For example,

$$64:\ \texttt{MOVE \#1}, \texttt{d}_0$$

which puts 1 into the data register $\texttt{d}_0$, is translated into

$$\forall t.\ \ \texttt{pc}(t) = 64 \rightarrow$$
$$\texttt{pc}(t + 1) = 66\, \wedge$$
$$\wedge\, \texttt{d}_0(t + 1) = 1 \wedge \texttt{d}_1(t + 1) = \texttt{d}_1(t) \wedge \ldots$$
$$\ldots \wedge \texttt{d}_7(t + 1) = \texttt{d}_7(t)\, \wedge$$
$$\wedge\, \texttt{a}_0(t + 1) = \texttt{a}_0(t) \wedge \ldots \wedge \texttt{a}_7(t + 1) = \texttt{a}_7(t)\, \wedge$$
$$\wedge\, \neg\texttt{Vflag}(t + 1) \wedge \neg\texttt{Cflag}(t + 1)\, \wedge$$
$$\wedge\, (\texttt{Zflag}(t + 1) \leftrightarrow 1 = 0)\, \wedge$$
$$\wedge\, (\texttt{Nflag}(t + 1) \leftrightarrow 1 < 0) \wedge \neg\texttt{Xflag}(t + 1)\, \wedge$$
$$\wedge\, \forall a.\, \texttt{memory}(a, t + 1) = \texttt{memory}(a, t)\ .$$

Also,

$$72\colon \text{BEQ } 8$$

which is spelt as "branch on equal", incrementing the program counter by 8 if the Z flag is set, is translated into

$$\forall t.\, \text{pc}(t) = 72 \rightarrow \; (\text{Zflag}(t) \rightarrow \text{pc}(t+1) = \text{pc}(t) + 8) \wedge$$
$$\wedge (\neg \text{Zflag}(t) \rightarrow \text{pc}(t+1) = \text{pc}(t) + 2) \wedge$$
$$\wedge \, d_0(t+1) = d_0(t) \wedge \ldots$$
$$\ldots \wedge d_7(t+1) = d_7(t) \wedge$$
$$\wedge \, a_0(t+1) = a_0(t) \wedge \ldots$$
$$\ldots \wedge a_7(t+1) = a_7(t) \wedge$$
$$\wedge \, (\text{Vflag}(t+1) \leftrightarrow \text{Vflag}(t)) \wedge$$
$$\wedge \, (\text{Zflag}(t+1) \leftrightarrow \text{Zflag}(t)) \wedge$$
$$\wedge \, (\text{Nflag}(t+1) \leftrightarrow \text{Nflag}(t)) \wedge$$
$$\wedge \, (\text{Cflag}(t+1) \leftrightarrow \text{Cflag}(t)) \wedge$$
$$\wedge \, (\text{Xflag}(t+1) \leftrightarrow \text{Xflag}(t)) \wedge$$
$$\wedge \, \forall a.\, \text{memory}(a, t+1) = \text{memory}(a, t) \; .$$

Although it may appear to the reader that this particular representation is very simple and verbose, we want to remark some points worth noticing:

- The theory of the microprocessor is an Harrop theory in **IL** (and a $\square$-theory in **E**).

- By a closer inspection, it results that the microprocessor theory as well as the output of the translation procedure, i.e., the program theory, are, indeed, closed specification frameworks.

- The verbose format of every instruction is, in fact, trivial, since it only depends on the instruction code (MOVE, BEQ in the examples) and on the addressing mode.

6.2 How to distinguish relevant information

In Chapter 5, we have seen that it is possible to extract information from a formal proof; we have remarked that the set of facts, i.e., $\text{Inf}^*(\mathcal{I})$, we extract, may be infinite. Of course, most of the content of $\text{Inf}^*(\mathcal{I})$ is useless from the point of view of the programmer; many facts are produced by the extraction algorithms because it needs enough information to be able to collect new proofs via the Exp operator.

There are two important points to put in evidence:

- the collection method is not intrinsically oriented. The procedure to extract information we described is not conceived to produce the minimal amount of facts in a proof that permits to redo the proof itself, as, for example, is the case of normalisation techniques [7, 8, 46].

- the notion of what is interesting depends on the user. We mean that, in our belief, is up to the user to choose what kind of information has to be considered as relevant. Moreover, we are not allowed to assume to know in advance, before doing a proof, what kind of information the user wants to extract.

Hence, the collection method is the right procedure to analyse correctness proofs, because it permits to *query* the proof for the kind of facts we consider as relevant, thus solving the second remark. Of course, we have to pay a price: being non oriented, the collection method can be very inefficient, that is, it may take a great number of steps to produce the information we are interested in.

We see the need for two implementation techniques: *orientation* and *filtering*. The former takes care of directing the collection method to try to extract the relevant information as soon as possible; the latter discriminates between interesting and non interesting facts. The idea behind the orientation techniques is to implement the algorithm constituting the collection method in a lazy functional style, that is, it has to compute step by step, producing a series of approximations $I_0, I_1, \ldots$ tending to $\mathrm{Inf}^*(\mathcal{I})$. As soon as we are able to ensure that $\bigcup_{i \in \omega} I_i = \mathrm{Inf}^*(\mathcal{I})$, any sequence of approximations $I_0, I_1, \ldots$ will be an instance of the collection method. Then, we can show to the user a sequence of facts $F_0, F_1, \ldots$ which is related to $I_0, I_1, \ldots$ by the formula $F_i = \tau(I_i)$, where τ is a filtering function that takes care of retaining only the part of I_i relevant for the purposes of the user.

The proposal we want to discuss here is a marking superstructure for proofs: some nodes in a proof tree are marked, and we treat in a privileged way the subproofs whose last inference step is marked. When we have to apply the basic operators Subproof, Coll, Inf and Exp to a set of proofs $\mathcal{I}$ with a marking superstructure $\mathcal{M}$, we choose to generate first the subproofs of the marked parts, and we collect the information giving maximum priority to the marked subproofs. We have also tried to use natural numbers as markers, with different combination of rules to *spread* the markers when running the extraction algorithm.

The results are encouraging, and we summarise them here:

- The marking superstructure, imposing an ordering on the subproofs, is *fair*, i.e., the approximating sets of facts it permits to extract from $\mathcal{I}$ tend to the limit $\text{Inf}^*(\mathcal{I})$.

- The sets the extraction algorithm produces first are more *significant*, i.e., contain more interesting information.

- It seems harder for this specialisation of the collection method to provide deeply hidden interesting information, that is, facts that are conclusions of subproofs generated by the Exp operator after a number of iterations. This result is not surprising since the Exp operator is effective when a large amount of facts is extracted in every phase, so to produce the possibility for unusual combinations of subproofs.

We experimented also another way of orienting the extraction procedure; we may divide a complete correctness proof Π into a set $L = \{\Pi_1, \ldots, \Pi_n\}$ of lemmata whose composition reconstitutes Π. In this way we really discard a part of the information we extract from Π; formally one may prove that $\text{Inf}^*(L) \subseteq \text{Inf}^*(\{\Pi\})$, but, in general, equality does not hold.

Hence, this technique, we call it *lemmification*, does not respect the idea of approximating $\text{Inf}^*(\{\Pi\})$ by an appropriate sequence of sets of facts, but, rather, it really discards parts which, a priori, we consider as non relevant. We adopted lemmification as the standard way to analyse proofs where some parts are proved by automatic decision procedures.

For example, if the proof $\begin{array}{c} \vdots \\ \phi \end{array}$ is divided into $L = \left\{ \begin{array}{cc} A & \\ \vdots\, 1, & \vdots\, 2 \\ \phi & A \end{array} \right\}$, where $\genfrac{}{}{0pt}{}{\vdots}{A}\,2$ has been proved by an automatic prover for Pres-

burger arithmetic[1], then we mark the proof $\genfrac{}{}{0pt}{}{A}{\vdots}{\phi}\,1$, thus we obtain

as a result that

1. First, we produce results about $\genfrac{}{}{0pt}{}{A}{\vdots}{\phi}\,1$, because it is marked,

   then about $\genfrac{}{}{0pt}{}{\vdots}{A}\,2$.

2. Information will never be collected examining instances

   of subproofs of $\genfrac{}{}{0pt}{}{\vdots}{\phi}$ which are neither subproofs of $\genfrac{}{}{0pt}{}{A}{\vdots}{\phi}\,1$ nor

   subproofs of $\genfrac{}{}{0pt}{}{\vdots}{A}\,2$.

At this point it should be clear to the reader the purpose of our techniques to orient the extraction procedure; on one side, we mark some parts to introduce a priority measure in the process; on the other side, we discard information that comes from the combination of unrelated proofs.

We know that these techniques are quite rough, and do not constitute a refined solution to the problem of orienting the information extraction process. The orienting problem for the collection method as well as for other information extraction procedures is, in fact, an open research problem.

The problem of filtering a sequence of sets of facts I_0, I_1, ..., to produce a similar sequence F_0, F_1, ..., where every F_i contains just the *relevant* elements of I_i, is of a different nature. In fact, the operation of filtering is very easy to describe from a technical point of view, but it requires a formal definition of the word *relevant*.

As we remarked in the beginning of this section, we do not want to fix once and forever what we consider as relevant, but, rather, to provide a flexible frame permitting to produce different kinds of information in a homogeneous way.

---

[1]Presburger arithmetic is Peano arithmetic with a very restricted multiplication. Its interest lies in the fact that it is decidable. So, it is very popular in automated theorem proving.

In our opinion, we must provide at least a tool that filters the facts which directly express what is valid in the program code. We developed this algorithm and we will show it in the following section. On the other side, a great amount of interesting facts are collected in the extraction process, and, yet, we have not a syntactical way to characterise them.

Again, as for the orientation techniques, we feel the need for a deeper series of theoretical results; without the ability to characterise classes of formulae by their role in a correctness proof, it appears to be very hard to describe in a formal, syntactical way relevant properties of programs we wish to filter out from the result Inf^* produces.

The labelling algorithm

As we anticipated in the previous discussion, we will now show a simple filtering algorithm that extracts information on what is true in specific points of an object code program.

We assume that the program P we are analysing is coded into the logical representation Rep_P, as previously described.

Definition 6.2.1 *Let* $\begin{array}{c}\text{Rep}_P\\ \vdots\\ \text{Spec}_P\end{array}$ *be a correctness proof for the program*

P, coded in a logical form as Rep_P, *with respect to the specification*

Spec_P; *let* $C = \text{Inf}^*\left(\left\{\begin{array}{c}\text{Rep}_P\\ \vdots\\ \text{Spec}_P\end{array}\right\}\right)$.

We define

$$\mathcal{L}_i = \{\exists x. A \mid (\exists x. \text{pc}(x) - i + 1 \wedge A) \in C\} \cup$$
$$\cup \{\forall x. A \mid (\forall x. \text{pc}(x) = i + 1 \to A) \in C\} \ ,$$

for every instruction, referred to by its position i inside the program code: moreover, we define $\mathcal{G} = \{\forall x. A \mid \forall x. A \in C\} \setminus \bigcup_i \mathcal{L}_i$.

Intuitively $\mathcal{L}_i$ is a set of assertions holding on the ith position (line) of the source code, while $\mathcal{G}$ contains a set of facts that are true everywhere in the program.

Consider the assembly code program in Figure 6.1: given a natural number in register d_0, it is divided by 2 and the result

```
1: MOVE # − 1, d₁
2: ADD    #1, d₁
3: MOVE  d₁, d₂
4: ADD   d₂, d₂
5: CMP   d₂, d₀
6: BGE   2
7: SUB   #1, d₁
```

Figure 6.1: Computing division by two.

$$\text{Rep} \equiv \text{pc}(0) = 1 \wedge 0 \leq d_0(0) \wedge (\forall t.\, I_1(t) \wedge \ldots I_8(t)) \wedge$$
$$\wedge (\forall t.\, \text{pc}(t) = 1 \vee \ldots \vee \text{pc}(t) = 8)$$
$$I_1(t) \equiv \text{pc}(t) = 1 \rightarrow \text{pc}(t+1) = 2 \wedge d_0(t+1) = d_0(t) \wedge$$
$$\wedge d_1(t+1) = -1$$
$$I_2(t) \equiv \text{pc}(t) = 2 \rightarrow \text{pc}(t+1) = 3 \wedge d_0(t+1) = d_0(t) \wedge$$
$$\wedge d_1(t+1) = 1 + d_1(t)$$
$$I_3(t) \equiv \text{pc}(t) = 3 \rightarrow \text{pc}(t+1) = 4 \wedge d_0(t+1) = d_0(t) \wedge$$
$$\wedge d_1(t+1) = d_1(t) \wedge$$
$$\wedge d_2(t+1) = d_1(t)$$
$$I_4(t) \equiv \text{pc}(t) = 4 \rightarrow \text{pc}(t+1) = 5 \wedge d_0(t+1) = d_0(t) \wedge$$
$$\wedge d_1(t+1) = d_1(t) \wedge$$
$$\wedge d_2(t+1) = 2d_2(t)$$
$$I_5(t) \equiv \text{pc}(t) = 5 \rightarrow \text{pc}(t+1) = 6 \wedge d_0(t+1) = d_0(t) \wedge$$
$$\wedge d_1(t+1) = d_1(t) \wedge$$
$$\wedge (N(t+1) \leftrightarrow d_1(t) \leq d_0(t))$$
$$I_6(t) \equiv \text{pc}(t) = 6 \rightarrow (N(t) \rightarrow \text{pc}(t+1) = 2) \wedge$$
$$\wedge (\neg N(t) \rightarrow \text{pc}(t+1) = 7) \wedge$$
$$\wedge d_0(t+1) = d_0(t) \wedge$$
$$\wedge d_1(t+1) = d_1(t)$$
$$I_7(t) \equiv \text{pc}(t) = 7 \rightarrow \text{pc}(t+1) = 8 \wedge d_0(t+1) = d_0(t) \wedge$$
$$\wedge d_1(t+1) = d_1(t) - 1$$
$$I_8(t) \equiv \text{pc}(t) = 8 \rightarrow \text{pc}(t+1) = 8 \wedge d_0(t+1) = d_0(t) \wedge$$
$$\wedge d_1(t+1) = d_1(t)$$

Figure 6.2: The logical representation for the program.

$$\begin{aligned} A \equiv\ & \mathrm{pc}(t) = 2 \rightarrow (2(1+\mathrm{d}_1(t)) \leq \mathrm{d}_0(t) \rightarrow \\ & \qquad\qquad \rightarrow \mathrm{pc}(t+5) = 2) \wedge \\ & \qquad\qquad \wedge (\mathrm{d}_0(t) < 2(1+\mathrm{d}_1(t)) \rightarrow \\ & \qquad\qquad\quad \rightarrow \mathrm{pc}(t+5) = 7) \wedge \\ & \qquad\qquad \wedge \mathrm{d}_1(t+5) = \mathrm{d}_1(t) \end{aligned}$$

$$B \equiv \forall\, t.\, \mathrm{d}_0(t) = \mathrm{d}_0(0)$$

$$C \equiv \exists\, t.\, \mathrm{pc}(t) = 2 \wedge \mathrm{d}_0(t) < 2(1+\mathrm{d}_1(t))$$

$$D \equiv \forall\, t.\, \mathrm{pc}(t) = 2 \rightarrow 2\mathrm{d}_1(t) \leq \mathrm{d}_0(t)$$

$$\dfrac{\dfrac{\dfrac{\text{Rep}}{A} \quad \dfrac{\text{Rep}}{B} \quad \dfrac{\text{Rep} \quad \dfrac{\text{Rep}}{A}}{C} \quad \dfrac{\text{Rep} \quad \dfrac{\text{Rep}}{B}}{D}}{\exists\, t.\, 2(\mathrm{d}_1(t)-1) \leq \mathrm{d}_0(0) \leq 2\mathrm{d}_1(t) - 1 \wedge \mathrm{pc}(t) = 7} \quad \text{Rep}}{\text{Spec}}\ \exists\mathrm{E}$$

Figure 6.3: The schema of the correctness proof.

is returned in register d_1. Formally:

$$\begin{aligned} \text{Spec} \equiv \exists\, t.\, \mathrm{pc}(t) = 8 \wedge (\mathrm{d}_0(0) &= 2\mathrm{d}_1(t) \vee \\ \vee\, \mathrm{d}_0(0) &= 2\mathrm{d}_1(t) + 1) \end{aligned}$$

The logical representation for this program is shown in Figure 6.2. The correctness proof for this program has been developed in the **E** logic: the essential schema for the correctness proof can be found in Figure 6.3.

Applying the collection method to this proof where the subproofs in the schema are considered as marked, and filtering the result according to Definition 6.2.1, we get that the set of facts true in the whole program and relevant for its correctness are

$$\mathcal{G} = \{\forall\, t.\, \mathrm{d}_0(t) = \mathrm{d}_0(0)\}$$

$: \mathcal{L}_0 = \{0 \le d_0(0)\}$

1: MOVE # $-1, d_1$

$: \mathcal{L}_1 = \{\exists t. d_0(t) < 2(d_1(t) + 1), \forall t. 2d_1(t) < d_0(t)\}$

2: ADD   #1, d_1
3: MOVE d_1, d_2
4: ADD d_2, d_2
5: CMP d_2, d_0
6: BGE   2

$: \mathcal{L}_6 = \{\exists t. 2(d_1(t) - 1) \le d_0(0) \le 2d_1(t) - 1\}$

7: SUB   #1, d_1

$: \mathcal{L}_7 = \{\exists t. d_0(0) = 2d_1(t) \lor d_0(0) = 2d_1(t) + 1\}$

Figure 6.4: The final output of the labelling algorithm.

while the labels indexed by the instruction number are

$$\mathcal{L}_0 = \{0 \le d_0(0)\}$$
$$\mathcal{L}_1 = \{\exists t. d_0(t) < 2(1 + d_1(t)), \forall t. 2d_1(t) < d_0(t)\}$$
$$\mathcal{L}_2 = \varnothing$$
$$\mathcal{L}_3 = \varnothing$$
$$\mathcal{L}_4 = \varnothing$$
$$\mathcal{L}_5 = \varnothing$$
$$\mathcal{L}_6 = \{\exists t. 2(d_1(t) - 1) \le d_0(0) \le 2d_1(t) - 1\}$$
$$\mathcal{L}_7 = \{\text{Spec}\}$$

Thus, we can automatically label the source code, obtaining the commented program in Figure 6.4.

The labelling algorithm is a first step in the direction of reflecting information from a correctness proof onto the originating program: it is works on a correctness proof interpreting some of its formulae in a special way; in particular, it is aware of the meaning of the pc function.

It is not difficult to develop variations over the labelling algorithm to filter different kinds of information. For example, if we filter the increment to the program counter, we get

$$\mathcal{L}_0 = \{1\}$$
$$\mathcal{L}_1 = \{2\}$$
$$\mathcal{L}_2 = \{3, 6, 11, \ldots\}$$
$$\mathcal{L}_3 = \{4, 7, 12, \ldots\}$$
$$\mathcal{L}_4 = \{5, 8, 13, \ldots\}$$
$$\mathcal{L}_5 = \{6, 9, 14, \ldots\}$$
$$\mathcal{L}_6 = \{x\}$$
$$\mathcal{L}_7 = \{x + 1\}$$

where x depends on the value of $d_0(0)$ and $\mathcal{L}_i$ represents the time after the execution of the instruction i.

This kind of information can be generated automatically searching for the witnesses of the existential labels in Figure 6.4. Obviously, closed forms for the equations describing the value of time after each instruction will be collected, if they were used in the correctness proof, leading to an explicit complexity evaluation.

Chapter 7

Uniformly Constructive
Formal Systems

The goal of this chapter is to define the notion of constructive system entailed by the collection method shown in Chapter 5.

The idea is to abstract over the algorithmic aspects of the method in order to build a purely mathematical definition of constructive system that allows a generalised notion of information extraction that can be still regarded as a "construction" in the usual sense of constructive logics.

The notion of constructivity we will develop allows us to define what we intend by *uniformly constructive systems*. This notion is based on the pragmatical principle that identifies the constructive systems with the formal systems allowing an effective information extraction procedure that captures the significant part of the information content of proofs, allowing in this way to interpret formulae and proofs of a formal system both in a logical and in a computational way. The content of this chapter is a simplification of [12, 15, 14, 17, 13].

7.1 Proofs and calculi

Let us assume to work within a (first-order) formal system whose language is $\mathcal{L}$.

Definition 7.1.1 *Given a formula A in the language $\mathcal{L}$, its degree* *is defined as:*

- $dg(A) = 1$ *if A is atomic[1];*

- $dg(A) = 1 + dg(B)$ *if $A \equiv \neg B$;*

- $dg(A) = 1 + \max\{dg(B), dg(C)\}$ *if $A \equiv B \wedge C$, $A \equiv B \vee C$ or $A \equiv B \rightarrow C$;*

- $dg(A) = 1 + dg(B(x))$ *if $A \equiv \forall x. B(x)$ or $A \equiv \exists x. B(x)$.*

The degree of a finite set of formulae Γ is defined to be $dg(\Gamma) = \max\{dg(\gamma) \mid \gamma \in \Gamma\}$.

A *proof* on the language $\mathcal{L}$ is a finite object π such that

- the finite set of formulae occurring in π, notation $\mathcal{F}(\pi)$, is uniquely determined and non empty;

- the sequent $Seq(\pi) = \Gamma \vdash \Delta$ proved by π is uniquely determined, where Γ and Δ are finite sets of formulae; Γ (possibly empty) is the set of *assumptions*, while Δ, that must be non empty, is the set of *conclusions* of π. The compact notation $\pi \colon \Gamma \vdash \Delta$ will be used to indicate that $Seq(\pi) = \Gamma \vdash \Delta$.

The degree of a proof, $dg(\pi)$ is defined as the maximum of the degrees of the formulae occurring in π; moreover the degree of a sequent, $dg(\Gamma \vdash \Delta)$ is defined to be the maximum among the degrees of the formulae occurring in $\Gamma \cup \Delta$.

Proofs are organised in calculi, defined as follows

Definition 7.1.2 *A calculus over $\mathcal{L}$ is a pair $\mathbf{C} = \langle C, [\cdot] \rangle$, where C is a recursive set of proofs in $\mathcal{L}$ and $[\cdot]$ is a recursive map from C into the set of finite subsets of C such that*

- $\pi \in [\pi]$;

- *for any $\pi' \in [\pi]$, $[\pi'] \subseteq [\pi]$;*

- *for any $\pi' \in [\pi]$, $dg(\pi') \leq dg(\pi)$.*

---

[1]In the case of the **E** logic, we consider as atomic w.r.t. the definition of degree also every boxed formula. In general, a formula is atomic when it is not decomposable w.r.t. to the constructive content of the logical connectives.

To lighten notation, we will identify a calculus $\mathbb{C}$ with C, the set of its proofs.

In the above definition, the $[\cdot]$ map denotes the set of *relevant* subproofs of a proof.

It is immediate to see that the set of proofs generated by the **IL** calculus as defined in Chapter 2, coupled with the map that associates to a proof the set of its subproofs as defined in Chapter 5 generates a calculus in the sense of Definition 7.1.2.

Analogously, defining C as the set of proofs generated by the inference rules for **E**, as shown in Chapter 2, and constructing the $[\cdot]$ map as the function associating to a proof the set of its subproofs, as defined in Chapter 5, it is immediate to show that **E** is a calculus in the sense of Definition 7.1.2.

Now, given a set of proofs $S \subseteq C$ in a fixed calculus $\mathbb{C}$, we denote with $[S]$ the *closure under subproofs* of S, namely, $[S] = \{\pi' \mid \exists \pi \in S. \pi' \in [\pi]\}$. In general, $[S]$ is not recursive, but, when S is finite, $[S]$ is obviously recursive, and hence $\langle [S], [\cdot]_{[S]} \rangle$ is a calculus, where $[\cdot]_{[S]}$ is the restriction of the map $[\cdot]$ of $\mathbb{C}$ to $[S]$.

Given a calculus $\mathbb{C}$, let $S \subseteq \mathbb{C}$:

- $\mathrm{Seq}(S) = \bigcup_{\pi \in S} \mathrm{Seq}(\pi)$;

- $\mathcal{F}(S) = \bigcup_{\pi \in S} \mathcal{F}(\pi)$;

- $\mathrm{dg}(S) = \max\{\mathrm{dg}(\pi) \mid \pi \in S\}$, where $\mathrm{dg}(S) = \infty$ if, for any number k, it is possible to find a proof in S whose degree exceeds k;

- $\mathcal{T}(S) = \{A \mid \vdash A \in \mathrm{Seq}(S)\}$, representing the set of *theorems* in S, i.e., the set of formulae proved in S.

Given a set $\mathcal{I}$ of formulae, representing the true facts in some formal system $\mathbb{S}$, we say that the calculus $\mathbb{C}$ is a *presentation* for $\mathcal{I}$ iff $\mathcal{T}(\mathbb{C}) = \mathcal{I}$.

In this sense, the **IL** calculus is a presentation of first-order intuitionistic logic, defined as the set of formulae that are true in every Kripke model as defined in Chapter 2.

7.2 Extraction rules

Fixed a calculus $\mathbb{C}$, we want to extract information from a set S of proofs by a sound manipulation of the subproofs of S, that is, every fact we may extract from S must be provable in $\mathbb{C}$, and, moreover, we must know how to prove it, by means of a suitable composition of the subproofs of S.

The notion of extraction rule wants to capture what is an admissible manipulation.

Definition 7.2.1 *An* extraction rule *(e-rule for short), is an inference rule of the form*

$$\frac{\Gamma_1 \vdash A_1 \quad \ldots \quad \Gamma_n \vdash A_n}{\Delta \vdash B} \; \mathcal{R}$$

where Γ_i, $1 \leq i \leq n$, and Δ are finite sets of formulae and A_i, $1 \leq i \leq n$, and B are formulae, such that

- *the e-rule $\mathcal{R}$ can be* uniformly simulated *in $\mathbb{C}$, that is, there exists a function $\phi \colon \mathbb{N} \to \mathbb{N}$ such that, for every $\pi_1 \colon \Gamma_1 \vdash A_1, \ldots, \pi_n \colon \Gamma_n \vdash A_n$ in $\mathbb{C}$, there is a proof $\pi \colon \Delta \vdash B$ in $\mathbb{C}$ such that $\mathrm{dg}(\pi) \leq \max\{\phi(\mathrm{dg}(\pi_1)), \ldots, \phi(\mathrm{dg}(\pi_n))\}$.*

- *If $n = 0$ ($\mathcal{R}$ is a zero-premises e-rule), then there exists $h \in \mathbb{N}$ such that $\mathrm{dg}(\Delta \vdash B) \leq h$.*

- *If $n > 0$ then $\mathcal{R}$ is* non-increasing, *that is,*

$$\mathrm{dg}(\Delta \vdash B) \leq \max\{\mathrm{dg}(\Gamma_1 \vdash A_1)), \ldots, \mathrm{dg}(\Gamma_n \vdash A_n))\} \; .$$

The first condition in the above definition says that the e-rule $\mathcal{R}$ is an admissible rule in the calculus $\mathbb{C}$, that is, it preserves the set of deducible sequents. Moreover, it says that $\mathcal{R}$ can be simulated by a proof whose complexity (represented by its degree) is bounded.

Definition 7.2.2 *A set $\mathcal{R}$ of e-rules for $\mathbb{C}$ is* h-bounded *($h \in \mathbb{N}$) if, for every zero-premises rule $R \equiv \dfrac{}{\Delta \vdash B} \in \mathcal{R}$, $\mathrm{dg}(\Delta \vdash B) \leq h$.*

Moreover, if $\mathcal{R} = \{R_1, \ldots, R_m\}$, where R_i is uniformly simulated in $\mathbb{C}$ w.r.t. ϕ_{R_i}, the whole set $\mathcal{R}$ is uniformly simulated w.r.t. $\phi_{\mathcal{R}} \colon \mathbb{N} \to \mathbb{N}$ defined as

$$\phi_{\mathcal{R}}(0) = \max\{\phi_{R_1}(0), \ldots, \phi_{R_m}(0)\}$$
$$\phi_{\mathcal{R}}(i+1) = \max\{\phi_{\mathcal{R}}(i), \phi_{R_1}(i+1), \ldots, \phi_{R_m}(i+1)\} \; .$$

$$\frac{\Gamma \vdash A}{\theta \Gamma \vdash \theta A} \text{ Subst} \qquad \frac{\Gamma \vdash A \quad \Delta, A \vdash B}{\Gamma, \Delta \vdash B} \text{ Cut}$$

$$\frac{}{\vdash x = x} \text{ Id}_1 \qquad \frac{\vdash A(t) \quad \vdash t = s}{\vdash A(s)} \text{ Id}_2$$

Figure 7.1: Examples of extraction rules

Thus, we can say that a finite set of extraction rules is h-bounded, or that it can be uniformly simulated in $\mathbb{C}$ by extending the conditions on the single rules.

Examples of extraction rules for **IL** + **ID**, i.e., intuitionistic logic with identity, are shown in Figure 7.1. It is straightforward to check that these are, indeed, e-rules, since Id_1 is a zero-premises rule, whose consequence is an atomic formula, thus it is 1-bounded; the other rules are obviously non-increasing. All these rules can be easily simulated in **IL** + **ID**.

Definition 7.2.3 *Given a h-bounded recursive set $\mathcal{R}$ of e-rules for $\mathbb{C}$ and a recursive set S of proofs of $\mathbb{C}$, we define the* extraction calculus *for $\mathbb{C}$ as the calculus $\mathbb{C}(\mathcal{R}, [S])$ having as axioms the sequents in $\text{Seq}([S])$ and as inference rules the set $\mathcal{R}$.*

In [12, 17] the following theorem has been proved

Theorem 7.2.1 *Let $\mathcal{R}$ be a h-bounded set of e-rules for $\mathbb{C}$ and let $S \subseteq \mathbb{C}$ with $\text{dg}(S) \leq k$, where $k \in \mathbb{N}$. Then:*

- *for every proof π in $\mathbb{C}(\mathcal{R}, [S])$, $\text{dg}(\pi) \leq \max\{h, k\}$;*

- *there is $S' \subseteq \mathbb{C}$ s.t. $\text{dg}(S') \leq \max\{k, \phi_{\mathcal{R}}(\max\{h, k\})\}$ and $\text{Seq}(S') = \text{Seq}(\mathbb{C}(\mathcal{R}, [S]))$.*

The first consequence of Theorem 7.2.1 says that the proofs generated by the extraction calculus are, at most, as complex as the proofs in S and the axioms in $\mathcal{R}$.

The second conclusion of Theorem 7.2.1 says that whatever formula A deduced from a proof in the extraction calculus can be deduced in the original calculus and, moreover, the proof of A in $\mathbb{C}$ has a bounded complexity.

7.3 Uniformly constructive formal systems

The notion of uniformly constructive system wants to discriminate between formal naïvely constructive systems that allows an effective information extraction procedure and the others.

A formal system is constructive in a naïve sense if

- whenever $\vdash A \vee B$, either $\vdash A$ or $\vdash B$;

- whenever $\vdash \exists x. A(x)$, there is a t such that $\vdash A(t)$.

This notion of constructive system is too limited, as discussed in, e.g., [49].

We distinguish two kinds of uniformly constructive systems; the criterion behind their distinction lies in the naïve requirement on existential statements: in fact, whenever $\vdash \exists x. A(x)$, we can require either the existence of a term t such that $\vdash A(t)$, or the existence of a *closed* term t such that $\vdash A(t)$. The former case is better suited for purely logical systems, such as **IL** or **E**, while the latter is better suited for logical theories, such as arithmetic, where the knowledge of a closed term permits the identification of an element in the intended domain that satisfies the existential statement.

Formally, we say that

Definition 7.3.1 *Let Γ be a set formulae, then Γ is (naïvely) open constructive iff*

- *if $A \vee B \in \Gamma$ then either $A \in \Gamma$ or $B \in \Gamma$;*

- *if $\exists x. A(x) \in \Gamma$ then $A(t) \in \Gamma$ for some t in the language.*

A set Γ is said to be (naïvely) closed constructive iff

- *if the closed formula $A \vee B \in \Gamma$ then either $A \in \Gamma$ or $B \in \Gamma$;*

- *if the closed formula $\exists x. A(x) \in \Gamma$ then $A(t) \in \Gamma$ for some closed term t in the language.*

A formal system is said to be uniformly constructive if it is open or closed constructive and it admits a bounded extraction calculus.

Definition 7.3.2 *Given a calculus $\mathbb{C} = \langle C, [\cdot] \rangle$, we say that $\mathbb{C}$ is uniformly constructive if there exists a finite h-bounded set $\mathcal{R}$ of* *e-rules for* C *such that, for every* $S \subseteq C$, *the set* $\{A \mid \; \vdash A \in \mathrm{Seq}(C(\mathcal{R}, [S]))\}$ *is open constructive.*

Analogously, we say that C *is* uniformly r-constructive *if there exists a finite h-bounded set* $\mathcal{R}$ *of e-rules for* C *such that, for every* $S \subseteq C$, *the set* $\{A \mid \vdash A \in \mathrm{Seq}(C(\mathcal{R}, [S]))\}$ *is closed constructive.*

The main consequence of these definitions comes from Theorem 7.2.1. In fact, if we prove $\pi \colon \; \vdash \exists x. A(x)$ in the uniformly constructive calculus C, then we can complete the information contained in the proof π by means of the extraction calculus $C(\mathcal{R}, [\pi])$. Being constructive, the completion process is assured to find a term t such that $\vdash A(t)$. The completing information can be found in the extraction calculus by means of an enumerative procedure involving only formulae of bounded complexity, e.g., the collection method.

It is possible to prove that a wide family of systems $T + L$ where T is a mathematical theory and L a logical system are uniformly constructive. In [12, 17] it has been shown that, if T is any Harrop theory and L is **IL**, the resulting formal system is uniformly constructive. Moreover, if we add to **IL** one of the following principles, the system is again uniformly constructive:

- the Grzegorczyk principle

$$(\forall x. A(x) \lor B) \to B \lor \forall x. A(x)$$

  with x not free of B;

- the Kuroda principle

$$\forall x. \neg\neg A(x) \to \neg\neg \forall a. A(x) \; ;$$

- the Extended Scott principle

$$((\forall x. \neg\neg A(x) \to A(x)) \to (\exists x. A(x) \lor \neg A(x))) \to$$
$$\to (\exists x. \neg A(x) \lor \neg\neg A(x)) \; ;$$

- the Kreisel-Putnam principle

$$(\neg A \to B \lor C) \to (\neg A \to B) \lor (\neg A \to C) \; ;$$

- the Independence of Premises principle

$$(\neg A \rightarrow \exists\, x.\, B(x)) \rightarrow (\exists\, x.\, \neg A \rightarrow B(x))$$

with x not free in A.

Similarly, in [12, 17], it has been shown that the theories formalising abstract datatypes according to the isoinitial approach plus the following mathematical principles are uniformly r-constructive:

- the Descending Chain principle

$$\exists\, x.\, A(x) \wedge (\forall\, y.\, A(y) \rightarrow (\exists\, z.\, A(z) \wedge z < y) \vee B) \rightarrow B\ ,$$

where $<$ is a well-founded ordering relation;

- the Markov principle

$$(\forall\, x.\, A(x) \vee \neg A(x)) \wedge \neg\neg \exists\, x.\, A(x) \rightarrow \exists\, x.\, A(x)\ ;$$

- the Transfinite Induction principle

$$\forall\, x.\, (\forall\, y.\, y < x \rightarrow A(y)) \rightarrow A(x) \rightarrow \forall\, z.\, A(z)$$

where $<$ is a well-founded ordering relation.

7.4 The collection method

We conclude this chapter with a discussion of the relation between the collection method and the notion of uniformly constructive system. The technique used to prove that a system $\mathbb{C}$ is uniformly constructive, see [12], proceeds as follows:

1. a set $\mathcal{R}$ of e-rules is selected in such a way that $\mathcal{R}$ is $\mathbb{C}$-closed, i.e., every conclusion we can obtain in the extraction calculus can be equivalently derived in $\mathbb{C}$ by means of a proof of bounded complexity;

2. a notion of evaluation on formulae is given, like Definition 5.1.9, which captures what means for a formula to be locally explained by a set of formulae;

3. for any proof $\pi\colon \Gamma \vdash A$ in $\mathbb{C}(\mathcal{R}, [S])$, a fixed the extraction calculus, with S a finite set of proofs of $\mathbb{C}$, we have to show that, if Γ is evaluated on the set of conclusions of S, then A is evaluated on the same set, as in Lemma 5.1.2;

4. one concludes that $\mathbb{C}(\mathcal{R}, [S])$ is open (closed) constructive, thus, by definition, $\mathbb{C}$ is uniformly (r-)constructive.

Thus, the basic lemma showing the closure on the notion of evaluation is fundamental in the proof that a system is uniformly constructive, as well as in the proof that a formal system admits a well-behaved collection procedure.

The first step above proves that a set of e-rules is admissible for a calculus. In the collection method this step correspond to the construction of the set Coll^*.

In fact, the collection method applied to **IL** operates as:

- given a set of proofs S, it combines proofs already in S by means of the CUT e-rule; this is done by means of the Coll operator.

- given a set of proofs S, it instantiates the eigenvariables in the subproofs of a parametric proof, by means of the SUBST e-rule; this is done by the Exp operator.

The above construction is iterated as far as possible until no new proofs can be generated. This fact corresponds to consider the whole set of proofs in the extraction calculus $\textbf{IL}(\{\text{CUT}, \text{SUBST}\}, [S])$.

Thus, proving the collection method generates a pseudo-truth set is a way to prove that the corresponding extraction calculus $\textbf{IL}(\{\text{CUT}, \text{SUBST}\}, [S])$ is open constructive. Being CUT and SUBST e-rules for **IL**, forming an h-bounded set, we conclude that **IL** is, indeed, uniformly constructive.

7.5 Concluding remarks

A natural question is "there are formal systems that are open (closed) constructive but not uniformly (r-)constructive?".

In other words, given an open (closed) constructive system, it is always possible to find an extraction method, e.g., an instance of the collection method, in order to give a constructive as well as a computational reading to its theorems?

The answer is NO: in [12, 17] a pathological formal system has been shown, HA*, that is closed constructive, but it cannot be uniformly r-constructive. It is based on an extension of Peano Arithmetic plus a non-obvious Gödelisation of formulae and proofs to force the impossibility to find an h-bounded set of uniform e-rules.

Another natural question is "it is always possible to prove by means of the collection method that a formal system is uniformly (r-)constructive?".

Again, the answer is NO: in [12] the Kreisel-Putnam logic and the Scott logic have been shown to be uniformly constructive, but, inspecting the proofs, one understands that the used notions of evaluation are incompatible with the structure of the collection method. The problem lies in the fact that the collection method requires to generate the Inf* set by composition of a set of proofs consisting of the original proofs plus the proofs constructed via the Exp operator. Usually, as in the case of **IL** and **E**, this order is not influent for the properties of the method, but limits the number of generated proofs. In the case of the Kreisel-Putnam logic, the order matters, thus the collection method fails to extract enough information, even if it is available.

Chapter 8

An Application of Program Analysis

This last chapter illustrates an application of the constructive reasoning about programs as depicted in the previous parts.

Instead of using the collection method or to build an extraction calculus that manipulates a correctness proof in order to extract relevant information, we will show how to decorate a correctness proof with appropriate labels allowing the extraction of significant information.

In particular, we discuss how to evaluate the temporal behaviour of a combinatorial circuit. Specifically, we assume to have a combinatorial circuit, i.e., a circuit with no memory, whose input signals are temporised, and we want to prove its correctness w.r.t. a given specification, and, moreover, we want to calculate the timing of the output signals, that is, when they become stable. More specifically, we show how the temporal information regarding the signals in the circuit can be extracted from the correctness proof. To this aim, we define an ad-hoc semantics of formulae in the E logic that carries not only the logical truth, but also the temporal information. Then, a suitable decoration procedure is designed which calculates the temporal information by analysing the structure of the correctness proof.

The discussed approach make a substantial use of the constructive nature of the E logic, since, as proved in the last section of this notes, the same result cannot be obtained in the domain of classical logic. Moreover, the notion of constructive negation plays a crucial role in the modelling of circuits, and it is not possible to transfer the same result in **IL** because the intuitionistic negation is, in a sense, too *poor* to model the circuits' behaviour in the described way.

The results of this lecture have been adapted from [16], and the interested reader is referred to that work for the missing details an for an in-depth discussion of the problem.

8.1 Waveforms and circuits

In the logical approach to circuit analysis a semantics represents an abstraction from the physical details and takes into account only the relevant aspects.

For example, let us consider the gates INV and NAND of Figure 8.1; their behaviour is specified by the following formulae in **CL**

$$INV(x,y) \equiv (x \rightarrow \neg y) \wedge (\neg x \rightarrow y) \qquad (8.1)$$
$$NAND(x,y,z) \equiv (x \wedge y \rightarrow \neg z) \wedge (\neg x \vee \neg y \rightarrow z) \quad (8.2)$$

The truth table of $INV(x,y)$ defines the input/output behaviour of the INV gate assuming x as input and y as output. Analogously, $NAND(x,y,z)$ represents the NAND gate, where x and y are the inputs and z is the output.

Similarly, the classical behaviour of the XOR circuit is specified by the formula

$$XOR(x,y,z) \equiv ((x \wedge \neg y) \vee (\neg x \wedge y) \rightarrow z) \wedge$$
$$\wedge ((x \wedge y) \vee (\neg x \wedge \neg y) \rightarrow \neg z)$$

where x and y represent the inputs and z the output.

Classical semantics allows us to study the input/output behaviour of combinatorial circuits, but prevents us to represent temporal information about the stabilisation properties of the circuits. Indeed, a more realistic description of the XOR circuit of Figure 8.1 should consider the instant at which the signals become stable and the delays in the propagation of signals, e.g., an "informal" characterisation of the behaviour of the above circuit should be as follows:

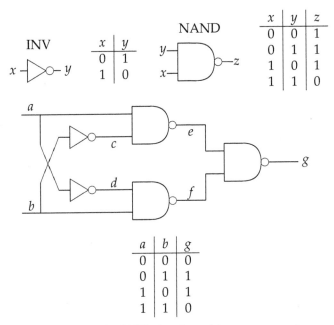

Figure 8.1: The XOR circuit and its components

$$
\boxed{
\begin{array}{c}
(a \text{ stable to } 1 \text{ at } t_1) \wedge (b \text{ stable to } 0 \text{ at } t_2) \\
\vee \\
(a \text{ stable to } 0 \text{ at } t_1) \wedge (b \text{ stable to } 1 \text{ at } t_2)
\end{array}
}
\longrightarrow
\begin{array}{c}
(g \text{ stable to } 1 \\
\text{at } F(t_1, t_2))
\end{array}
$$

$$
\boxed{
\begin{array}{c}
(a \text{ stable to } 1 \text{ at } t_1) \wedge (b \text{ stable to } 1 \text{ at } t_2) \\
\vee \\
(a \text{ stable to } 0 \text{ at } t_1) \wedge (b \text{ stable to } 0 \text{ at } t_2)
\end{array}
}
\longrightarrow
\begin{array}{c}
(g \text{ stable to } 0 \\
\text{at } G(t_1, t_2))
\end{array}
$$

where F and G are some functions $\mathbb{N}^2 \to \mathbb{N}$ and $\mathbb{N}$ represents discrete time.

Following [29, 28], a *signal* is a discrete timed boolean function $\sigma\colon \mathbb{N} \to \mathbb{B}$. A *circuit* is characterised by a set S of *observables* (the atomic formulae of our language) and a *waveform* is a map $V\colon S \to (\mathbb{N} \to \mathbb{B})$ associating with every observable a signal. A waveform represents an observable property of a circuit C, and an observable *behaviour* of C is described by a set of waveforms.

As an example, to represent the XOR circuit we need the set of observables $\{a, b, c, d, e, f, g\}$ representing the connections between the gates of the circuit. Thus, waveforms represent possible behaviours of the circuit.

Since we are interested in studying the stabilisation properties of a circuit, we consider only waveforms that stabilise at some time. In particular, we introduce the following notions of stabilisation for a waveform:

1. V is *stable* iff, for every $a \in S$ and for every $t \in \mathbb{N}$, $V(a)(t) = V(a)(0)$;

2. V is *eventually stable* iff, for every $a \in S$, there exists $t \in \mathbb{N}$ such that, for every $k \geq t$, $V(a)(k) = V(a)(t)$.

We denote the set of all the stable waveforms with STABLE and the set of all the eventually stable waveforms as EStable.

To express the *stabilisation properties* of waveforms and behaviours, we use a propositional language $\mathcal{L}_S$ based on the denumerable set S of *observables*. The formulae of $\mathcal{L}_S$ are the propositional formulae of $\mathbf{E}$ over the language $\mathcal{L}_S$.

A waveform V *validates a stabilisation property A at time t*, written $t, V \Vdash A$, if one of the following conditions holds:

1. $t, V \Vdash a$, where $a \in S$, iff $V(a)(k) = 1$ for all $k \geq t$;

2. $t, V \Vdash \Box B$ iff $k, V \Vdash B$ for some $k \geq t$;

3. $t, V \Vdash B \wedge C$ iff $t, V \Vdash B$ and $t, V \Vdash C$;

4. $t, V \Vdash B \vee C$ iff either $t, V \Vdash B$ or $t, V \Vdash C$;

5. $t, V \Vdash B \to C$ iff, for every $k \in \mathbb{N}$, $k, V \Vdash B$ implies $l, V \Vdash C$ for some $l \geq k$;

6. $t, V \Vdash \neg a$, where $a \in S$, iff $V(a)(k) = 0$ for all $k \geq t$;

7. $t, V \Vdash \neg \Box B$ iff $k, V \Vdash \neg B$ for some $k \geq t$;

8. $t, V \Vdash \neg(B \wedge C)$ iff either $t, V \Vdash \neg B$ or $t, V \Vdash \neg C$;

9. $t, V \Vdash \neg(B \vee C)$ iff $t, V \Vdash \neg B$ and $t, V \Vdash \neg C$;

10. $t, V \Vdash \neg(B \to C)$ iff $t, V \Vdash B$ and $t, V \Vdash \neg C$;

11. $t, V \Vdash \neg\neg B$ iff $t, V \Vdash B$.

It is immediate to check that $t, V \Vdash A$ implies $h, V \Vdash A$, for all $h \geq t$. For a atomic, a and $\neg a$ denote the stability of the observable signal $V(a)$ (at time t, with value 1 and 0, respectively). Indeed $t, V \Vdash a$ ($t, V \Vdash \neg a$) iff the signal $V(a)$ is stable to 1 (to 0 respectively) from t on.

Implication underlies a propagation delay, i.e., $t, V \Vdash A \to B$ means that, whenever, at some t', A "stabilises" ($t', V \Vdash A$) then, after a certain amount of time s, B will "stabilise" ($t' + s, V \Vdash B$).

We also remark that, differently from the other connectives, the validity of an implication is independent of t, indeed, $t, V \Vdash (B \to C)$ iff $0, V \Vdash (B \to C)$. Intuitively this corresponds to the fact that an implication does not represent a *property observable at a given time*, but a *global property* expressing a behaviour invariant with respect to time shifts.

Finally, the constructive understanding of negation is essential since $\neg a$ states the *positive* information "a stabilises to 0" and thus it differs from the usual intuitionistic understanding of negation as "a implies falsehood".

A logical characterisation of stable and eventually stable waveforms is:

$$V \in \text{STABLE iff } 0, V \Vdash A \vee \neg A \text{ for every } A \in \mathcal{L}_S ,$$
$$V \in \text{ESTABLE iff } 0, V \Vdash \Box A \vee \neg \Box A \text{ for every } A \in \mathcal{L}_S .$$

Now, to represent the classical input/output behaviour of a boolean function in our semantics, we associate to an eventually stable waveform V the classical interpretation $\langle V \rangle$ as follows: for every $a \in S$,

$$\langle V \rangle (a) = \begin{cases} 0 & \text{if} \quad 0, V \Vdash \neg \Box a , \\ 1 & \text{if} \quad 0, V \Vdash \Box a . \end{cases}$$

Definition 8.1.1 *A formula $F(a_1, \ldots, a_n, b)$ in the language $\mathcal{L}_S$ represents a boolean function $f : \mathbb{B}^n \to \mathbb{B}$ iff, for every $V \in$ ESTABLE, $0, V \Vdash \Box F(a_1, \ldots, a_n, b)$ iff $f(\langle V \rangle (a_1), \ldots, \langle V \rangle (a_n)) = \langle V \rangle (b)$.*

We remark that the above definition works when $0, V \Vdash \Box A \vee \neg \Box A$ holds, that is, when V is eventually stable.

The *formal verification task* of the circuit of Figure 8.1 consists in exhibiting a formal proof of the formula

$$\text{INV}(b,c) \wedge \text{INV}(a,d) \wedge \text{NAND}(a,c,e) \wedge$$
$$\wedge \text{NAND}(b,d,f) \wedge \text{NAND}(e,f,g) \rightarrow \text{XOR}(a,b,g) \ .$$

If our aim is only to prove the correctness of the above circuit **CL** is sufficient. But if we aim to extract information about the stabilisation delays of the circuit from the correctness proof, we need to introduce an intensional semantics of formulae that takes into account the temporal information.

8.2 Stabilisation bounds

The validation ⊩ provides an interpretation of formulae as stabilisation properties, but the information about stabilisation delays is not explicit. To extract stabilisation delays we need an analysis of all the waveforms of a behaviour. To deal with delays in our logic, we use the notion of *stabilisation bound* introduced in [27] and inspired by the *evaluation forms* of [38]. Evaluation forms correspond to structural truth evaluations of formulae; stabilisation bounds combine both truth and timing analysis.

Formally, we assign to every formula A of $\mathcal{L}_S$ a set of *stabilisation bounds* $\lceil A \rceil$ and an equivalence relation $\sim_A$ between elements of $\lceil A \rceil$, inductively defined on the structure of A:

- if $A = a$ or $A = \neg a$, with $a \in S$, then $\lceil A \rceil = \mathbb{N}$, and $t \sim_A t'$ for every $t, t' \in \lceil A \rceil$;

- if $A = \Box B$ or $A = \neg\Box B$ then $\lceil A \rceil = \{0\}$, and $0 \sim_A 0$;

- $\lceil B \wedge C \rceil = \lceil B \rceil \times \lceil C \rceil$ and $(\beta, \gamma) \sim_{B \wedge C} (\beta', \gamma')$ iff $\beta \sim_B \beta'$ and $\gamma \sim_C \gamma'$;

- $\lceil A_1 \vee A_2 \rceil = \lceil A_1 \rceil \oplus \lceil A_2 \rceil$ (where $\oplus$ denotes the disjoint sum) with $\alpha \in \lceil A_1 \rceil$ and $\alpha' \in \lceil A_2 \rceil$) and $(i, \alpha) \sim_{A_1 \vee A_2} (j, \alpha')$ iff $i = j$ and $\alpha \sim_{A_i} \alpha'$;

- $\lceil B \rightarrow C \rceil = \{f \mid f \colon \lceil B \rceil \rightarrow \lceil C \rceil \text{ such that } \beta \sim_B \beta' \text{ implies } f(\beta) \sim_C f(\beta')\}$, and $f \sim_{B \rightarrow C} f'$ iff $f(\beta) \sim_C f'(\beta)$ for every $\beta \in \lceil B \rceil$;

- $\lceil \neg(A_1 \wedge A_2) \rceil = \lceil \neg A_1 \rceil \oplus \lceil \neg A_2 \rceil$ and $(i, \alpha) \sim_{\neg(A_1 \wedge A_2)}$ (j, α') iff $i = j$ and $\alpha \sim_{\neg A_i} \alpha'$;

- $\lceil \neg(B \vee C) \rceil = \lceil \neg B \rceil \times \lceil \neg C \rceil$ and $(\beta, \gamma) \sim_{\neg(B \vee C)} (\beta', \gamma')$ iff $\beta \sim_{\neg B} \beta'$ and $\gamma \sim_{\neg C} \gamma'$;

- $\lceil \neg(B \rightarrow C) \rceil = \lceil B \rceil \times \lceil \neg C \rceil$ and $(\beta, \gamma) \sim_{\neg(B \rightarrow C)} (\beta', \gamma')$ iff $\beta \sim_B \beta'$ and $\gamma \sim_{\neg C} \gamma'$;

- $\lceil \neg\neg B \rceil = \lceil B \rceil$ and $\beta \sim_{\neg\neg B} \beta'$ iff $\beta \sim_B \beta'$.

The equivalence relation $\sim_A$ is needed to cut undesired functions in the definition of $\lceil B \rightarrow C \rceil$. Intuitively, a stabilisation bound $\alpha \in \lceil A \rceil$ intensionally represents a set of waveforms V that validate A for the "same reasons" and with the "same delay bounds". Formally, let us denote with V^t the waveform obtained by shifting V of t, i.e.,

$$V^t(a)(k) = V(a)(t + k) \text{ for all } a \in S, k \in \mathbb{N}$$

A waveform V *validates* A *with stabilisation bound* α, and we write $\alpha, V \models A$, if one of the following conditions holds:

1. $t, V \models a$, with $a \in S$, iff $t, V \Vdash a$;

2. $t, V \models \neg a$, with $a \in S$, iff $t, V \Vdash \neg a$;

3. $0, V \models \Box B$ iff $t, V \Vdash B$ for some $t \in \mathbb{N}$;

4. $(\beta, \gamma), V \models B \wedge C$ iff $\beta, V \models B$ and $\gamma, V \models C$;

5. $(i, \alpha), V \models A_1 \vee A_2$ iff $\alpha, V \models A_i$, where $i \in \{1, 2\}$;

6. $f, V \models B \rightarrow C$ iff, for every $t \in \mathbb{N}$ and $\beta \in \lceil B \rceil$, $\beta, V^t \models B$ implies $f(\beta), V^t \models C$;

7. $0, V \models \neg\Box B$ iff $t, V \Vdash \neg B$ for some $t \in \mathbb{N}$;

8. $(i, \alpha), V \models \neg(A_1 \wedge A_2)$ iff $\alpha, V \models \neg A_i$, where $i \in \{1, 2\}$;

9. $(\beta, \gamma), V \models \neg(B \vee C)$ iff $\beta, V \models \neg B$ and $\gamma, V \models \neg C$;

10. $(\beta, \gamma), V \models \neg(B \rightarrow C)$ iff $\beta, V \models B$ and $\gamma, V \models \neg C$;

11. $\beta, V \models \neg\neg B$ iff $\beta, V \models B$.

As an example, the INV and NAND gates have the following sets of stabilisation bounds:

$$\lceil INV(x,y) \rceil = (\mathbb{N} \to \mathbb{N}) \times (\mathbb{N} \to \mathbb{N}) \; ,$$
$$\lceil NAND(x,y,z) \rceil = (\mathbb{N} \times \mathbb{N} \to \mathbb{N}) \times (\mathbb{N} \oplus \mathbb{N} \to \mathbb{N}) \; .$$

A stabilisation bound for $INV(x,y)$ is, for example, the pair of identical functions (f_{INV}, f_{INV}) where

$$f_{INV}(t) = t + \delta_I \tag{8.3}$$

representing the set of valuations V such that $V(y)$ stabilises at time $t + \delta_I$ if $V(x)$ stabilises at time t with delay δ_I.

Analogously, $(f_{NAND}^-, f_{NAND}^+) \in \lceil NAND(x,y,z) \rceil$, with

$$\begin{aligned} f_{NAND}^-((t_1, t_2)) &= \max\{t_1, t_2\} + \delta_N \text{ and} \\ f_{NAND}^+((i, t)) &= t + \delta_N \end{aligned} \tag{8.4}$$

is an example of a data-independent stabilisation bound for the NAND gate. Indeed in f_{NAND}^-, δ_N is independent from t_1 (the time at which x stabilises to 1) and from t_2 (the time at which y stabilises to 1); analogously, in f_{NAND}^+, δ_N is independent of the pair (i, t).

We point out that, in general, stabilisation bounds represent data-dependent information; e.g., a stabilisation bound for $NAND(x,y,z)$ may consist of a pair of functions (η^-, η^+), where η^- calculates the stabilisation bound for output stable to 0 and η^+ for output stable to 1.

It is easy to prove that validity is preserved by time shifting, i.e., $\alpha, V \models A$ implies $\alpha, V^t \models A$ for every $t \in \mathbb{N}$.

Moreover, it is easy to check the following result:

Proposition 8.2.1 *Let T be the following time evaluation function:*

- $T(t) = t$, *for $t \in \mathbb{N}$;*

- $T((\alpha, \beta)) = \max\{T(\alpha), T(\beta)\}$;

- $T((i, \alpha)) = T(\alpha)$, *for $i = 1, 2$;*

- $T(f) = 0$, *with f any function.*

Let V be a waveform and let A be a formula. For every $t \in \mathbb{N}$, $t, V \Vdash A$ iff there is $\alpha \in \lceil A \rceil$ such that $T(\alpha) \leq t$ and $\alpha, V \models A$.

Proposition 8.2.1 links the intensional semantics based on stabilisation bounds and the extensional semantics defined in Section 8.1.

Let A be a formula, let $\alpha \in \lceil A \rceil$ and let V be a waveform; α is *exact for V and A* if $\alpha, V \models A$ and one of the following conditions holds:

1. $A = \Box B$ or $A = \neg\Box B$;

2. $A = a$ or $A = \neg a$, with $a \in S$, and, for all $t \in \mathbb{N}, t, V \models A$ implies $\alpha \leq t$;

3. $A = B \wedge C$, $\alpha = (\beta, \gamma)$, β is exact for B and V, and γ is exact for C and V;

4. $A = B_1 \vee B_2$, $\alpha = (k, \beta_k)$, with $k \in \{1, 2\}$, and β_k is exact for V and B_k;

5. $A = B \rightarrow C$ and, for all $\beta \in \lceil B \rceil$, if β is exact for V and B, then $\alpha(\beta)$ is exact for V and C;

6. $A = \neg\neg B$ and α is exact for V and B;

7. $A = \neg(B_1 \wedge B_2)$, $\alpha = (k, \beta_k)$, with $k \in \{1, 2\}$, and β_k is exact for V and $\neg B_k$;

8. $A = \neg(B \vee C)$, $\alpha = (\beta, \gamma)$, β is exact for V and $\neg B$, γ is exact for V and $\neg C$;

9. $A = \neg(B \rightarrow C)$, $\alpha = (\beta, \gamma)$, β is exact for V and B, γ is exact for V and $\neg C$.

Thus, α is exact for V and A iff α is the *exact* time when the computed signal A stabilises, with input the waveforms V.

8.3 Timing analysis of a circuit

Let us consider the problem to compute the stabilisation delays of the XOR circuit in Figure 8.1.

Firstly, we have to provide a complete description of the components of the circuit: for every component of the circuit, we have to provide a representing formula A and a time bound $\alpha \in \lceil A \rceil$ which is exact for the set V of observed behaviours (the set of waveforms resulting from an experimental analysis of the component).

In our example the description is given by the formulae:

- $INV(b,c)$ and $INV(a,d)$, obtained by instantiating the formula $INV(x,y)$ of (8.1);

- $NAND(a,c,e)$, $NAND(b,d,f)$ and $NAND(e,f,g)$, that are instances the formula $NAND(x,y,z)$ of (8.2).

As for the stabilisation bounds, we assume that:

- all the instances of the INV gate occurring in the circuit have the same stabilisation bound (f_{INV}, f_{INV}) described in (8.3);

- all the instances of the NAND gate occurring in the circuit have the same stabilisation bound (f_{NAND}^-, f_{NAND}^+) described in (8.4).

Starting from this information we want to compute an exact stabilisation bound for the whole circuit.

Calling $\mathcal{C}_{XOR}$ the description of the XOR circuit, if there exists a proof

$$\Pi : \mathcal{C}_{XOR} \vdash XOR(a,b,g)$$

in **E**, then, since $INV(x,y)$, $NAND(x,y,z)$ and $XOR(x,y,z)$ represent the corresponding boolean functions inv, nand and xor according to Definition 8.1.1, the input/output behaviour of the XOR circuit of Figure 8.1 is proved to be correct. Obviously, this holds also if Π is a proof of classical logic.

But, we are going to show that we can extract information about the stabilisation delays from the proof in **E**.

Computing stabilisation delays

To this aim, we associate with every proof $\pi : \{A_1, \dots, A_n\} \vdash B$ in **E**, a function $F_\pi : \lceil A_1 \rceil \times \cdots \times \lceil A_n \rceil \to \lceil B \rceil$: in the following, $\underline{\alpha}$ denotes an element of $\lceil A_1 \rceil \times \cdots \times \lceil A_n \rceil$. The function is defined by induction on the structure of the proof π:

- assumption: in this case, F_π is the identity function;

- $\wedge I, \neg \vee I, \neg \to I$: in these cases π has the form

$$\begin{array}{cc} A_1, \dots, A_k & A_{k+1}, \dots, A_n \\ \vdots\ \pi_1 & \vdots\ \pi_2 \\ C_1 & C_2 \\ \hline \multicolumn{2}{c}{B} \end{array}$$

and $F_\pi(\underline{\alpha}) = (F_{\pi_1}(\alpha_1, \ldots, \alpha_k), F_{\pi_2}(\alpha_{k+1}, \ldots, \alpha_n))$;

- $\wedge E, \neg \vee E, \neg \to E$: in these cases π has the form

$$
\begin{array}{c}
A_1, \ldots, A_n \\
\vdots\ \pi_1 \\
C \\
\hline
B
\end{array}
$$

and $F_\pi(\underline{\alpha}) = (F_{\pi_1}(\underline{\alpha}))_i$, where i denotes the first or the second element, according to the use of the first or the second instance of the rule in Table 2.2;

- $\vee I, \neg \wedge I$: in these cases π has the form

$$
\begin{array}{c}
A_1, \ldots, A_n \\
\vdots\ \pi_1 \\
C \\
\hline
B
\end{array}
$$

and $F_\pi(\underline{\alpha}) = (i, F_{\pi_1}(\underline{\alpha}))$, where i denotes the use of the first or the second instance of the rule in Table 2.2;

- $\vee E, \neg \wedge E$: in these cases π has the form

$$
\begin{array}{ccc}
A_1, \ldots, A_k & A_{k+1}, \ldots, A_l, [C_1] & A_{l+1}, \ldots, A_n, [C_2] \\
\vdots\ \pi_1 & \vdots\ \pi_2 & \vdots\ \pi_3 \\
B & B & B \\
\hline
& B &
\end{array}
$$

and

$$
F_\pi(\underline{\alpha}) = \begin{cases} F_{\pi_2}(\alpha_{k+1}, \ldots, \alpha_l, \beta), & \text{if } F_{\pi_1}(\alpha_1, \ldots, \alpha_k) = (1, \beta) \\ F_{\pi_3}(\alpha_{l+1}, \ldots, \alpha_n, \gamma), & \text{if } F_{\pi_1}(\alpha_1, \ldots, \alpha_k) = (2, \gamma); \end{cases}
$$

- $\to I$: in this case π is the proof

$$
\begin{array}{c}
A_1, \ldots, A_n, [B] \\
\vdots\ \pi_1 \\
C \\
\hline
B \to C
\end{array}
$$

and $F_\pi(\underline{\alpha})$ is the function $f : \lceil B \rceil \to \lceil C \rceil$ such that $f(\beta) = F_{\pi_1}(\underline{\alpha}, \beta)$;

- →E: in this case π is the proof

$$
\begin{array}{cc}
A_1, \ldots, A_k & A_{k+1}, \ldots, A_n \\
\vdots\ \pi_1 & \vdots\ \pi_2 \\
C & C \rightarrow B \\
\hline
\end{array}
$$
$$
B
$$

and $F_\pi(\underline{\alpha}) = F_{\pi_2}(\alpha_{k+1}, \ldots, \alpha_n)(F_{\pi_1}(\alpha_1, \ldots, \alpha_k))$;

- ¬E: in this case $F_\pi(\underline{\alpha}) = \gamma$ where γ is an arbitrary element in $\lceil B \rceil$;

- ¬¬I, ¬¬E: in these cases π has the form

$$
\begin{array}{c}
A_1, \ldots, A_n \\
\vdots\ \pi_1 \\
C \\
\hline
B
\end{array}
$$

and $F_\pi(\underline{\alpha}) = F_{\pi_1}(\underline{\alpha})$;

- □I, ¬□I: in these cases $F_\pi(\underline{\alpha}) = 0$.

The main properties of the function F_π associated with a proof $\pi \in \mathbf{E}$ are given by the following result.

Theorem 8.3.1 *Let* $\pi\colon \{A_1, \ldots, A_n\} \vdash B$ *be a proof in* $\mathbf{E}$ *and let* $F_\pi\colon \lceil A_1 \rceil \times \cdots \times \lceil A_n \rceil \rightarrow \lceil B \rceil$ *be the function associated with* π. *For all* $\alpha_1 \in \lceil A_1 \rceil, \ldots, \alpha_n \in \lceil A_n \rceil$, *and for every eventually stable waveform* V:

1. $\alpha_1, V \models A_1, \ldots, \alpha_n, V \models A_n$ *imply* $F_\pi(\alpha_1, \ldots, \alpha_n), V \models B$.

2. *if* $\alpha_1' \sim_{A_1} \alpha_1, \ldots, \alpha_n' \sim_{A_n} \alpha_n$, *then it holds the statement* $F_\pi(\alpha_1', \ldots, \alpha_n') \sim_B F_\pi(\alpha_1, \ldots, \alpha_n)$.

3. α_1 *exact for* V *and* $A_1, \ldots, \alpha_n$ *exact for* V *and* A_n *imply* $F_\pi(\alpha_1, \ldots, \alpha_n)$ *exact for* V *and* B.

Proof: By induction on the structure of the proof π: if π only consists of an assumption introduction (the base case), then F_π is the identity on $\lceil A \rceil$ and the assertions trivially follow; the induction step goes by cases according to the last rule applied in π. The details can be found in [16]. □

Application to the XOR circuit

In this subsection we apply Theorem 8.3.1 to compute the exact stabilisation bounds for the XOR circuit of Figure 8.1.

To this aim, firstly we describe the formal correctness proof $\Pi: C_{XOR} \vdash XOR(a, b, g)$ in E, then we show how to construct the function F_Π.

The proof can be constructed as follows:

$$\Pi \equiv \cfrac{\begin{matrix} \Gamma_3 \\ \vdots\; \pi_3 \\ (a \wedge \neg b) \vee (\neg a \wedge b) \to g \end{matrix} \qquad \begin{matrix} \Gamma_6 \\ \vdots\; \pi_6 \\ (a \wedge b) \vee (\neg a \wedge \neg b) \to \neg g \end{matrix}}{XOR(a, b, g)} \wedge I$$

where the proofs π_3 and π_6 are described below.

$$\pi_3 \equiv \cfrac{\cfrac{\left[\begin{matrix}(a \wedge \neg b) \vee \\ \vee (\neg a \wedge b)\end{matrix}\right] \quad \cfrac{\begin{matrix}[a \wedge \neg b], \; [\neg a \wedge b], \\ \Gamma_1 \qquad \Gamma_2 \\ \vdots\; \pi_1 \qquad \vdots\; \pi_2 \\ \neg e \vee \neg f \quad \neg e \vee \neg f\end{matrix}}{\neg e \vee \neg f} VE \quad \cfrac{NAND(e, f, g)}{\neg e \vee \neg f \to g} \wedge E}{g} \to E}{(a \wedge \neg b) \vee (\neg a \wedge b) \to g} \to I$$

where $\Gamma_3 = \Gamma_1 \cup \Gamma_2 \cup \{NAND(e, f, g)\}$.

$$\pi_6 \equiv \cfrac{\cfrac{\left[\begin{matrix}(a \wedge \neg b) \vee \\ \vee (\neg a \wedge b)\end{matrix}\right] \quad \cfrac{\begin{matrix}[a \wedge b], \; [\neg a \wedge \neg b], \\ \Gamma_4 \qquad \Gamma_5 \\ \vdots\; \pi_4 \qquad \vdots\; \pi_5 \\ e \wedge f \qquad e \wedge f\end{matrix}}{e \wedge f} VE \quad \cfrac{NAND(e, f, g)}{e \wedge f \to \neg g} \wedge E}{\neg g} \to E}{(a \wedge b) \vee (\neg a \wedge \neg b) \to \neg g} \to I$$

where $\Gamma_6 = \Gamma_4 \cup \Gamma_5 \cup \{NAND(e, f, g)\}$.

$$\pi_1 \equiv \dfrac{\dfrac{\dfrac{a \wedge \neg b}{a} \wedge\text{E} \qquad \dfrac{\dfrac{a \wedge \neg b}{\neg b} \wedge\text{E} \quad \dfrac{\text{INV}(b,c)}{\neg b \to c} \wedge\text{E}}{c} \to\text{E}}{a \wedge c} \wedge\text{I} \qquad \dfrac{\text{NAND}(a,c,e)}{a \wedge c \to \neg e} \wedge\text{E}}{\dfrac{\dfrac{\neg e}{} \to\text{E}}{\neg e \vee \neg f} \vee\text{I}}$$

where $\Gamma_1 = \{\text{INV}(b,c), \text{NAND}(a,c,e)\}$.

$$\pi_2 \equiv \dfrac{\dfrac{\dfrac{\neg a \wedge b}{b} \wedge\text{E} \qquad \dfrac{\dfrac{\neg a \wedge b}{\neg a} \wedge\text{E} \quad \dfrac{\text{INV}(a,d)}{\neg a \to d} \wedge\text{E}}{d} \to\text{E}}{b \wedge d} \wedge\text{I} \qquad \dfrac{\text{NAND}(b,d,f)}{b \wedge d \to \neg f} \wedge\text{E}}{\dfrac{\dfrac{\neg f}{} \to\text{E}}{\neg e \vee \neg f} \vee\text{I}}$$

where $\Gamma_2 = \{\text{INV}(a,d), \text{NAND}(b,d,f)\}$.

Now, the function associated with Π has the form:

$$F_\Pi : \quad \lceil\text{INV}(b,c)\rceil \times \lceil\text{INV}(a,d)\rceil \times \lceil\text{NAND}(a,c,e)\rceil \times$$
$$\times \lceil\text{NAND}(b,d,f)\rceil \times \lceil\text{NAND}(e,f,g)\rceil \to$$
$$\to \lceil\text{XOR}(a,b,g)\rceil$$

In general we can associate with every formula in $\mathcal{C}_{\text{XOR}}$ a different stabilisation bound, however, we assume that:

- All the instances of the formula $\text{INV}(x,y)$ have the same stabilisation bound $\iota = (\iota^-, \iota^+)$;

- All the instances of the formula $\text{NAND}(x,y,z)$ have the same stabilisation bound $\eta = (\eta^-, \eta^+)$.

With these assumptions, instead of $F_\Pi(\iota, \iota, \eta, \eta, \eta)$, we can write $F_\Pi(\iota, \eta)$. To construct the function F_π we follow the definition:

$$F_\Pi(\iota, \eta) = ((F_{\pi_3}(\iota, \eta), F_{\pi_6}(\iota, \eta))) \in (\mathbb{N}^2 \oplus \mathbb{N}^2 \to \mathbb{N})^2$$

where F_{π_3} and F_{π_6} are the functions associated with the subproofs π_3 and π_6. The construction goes on as follows:

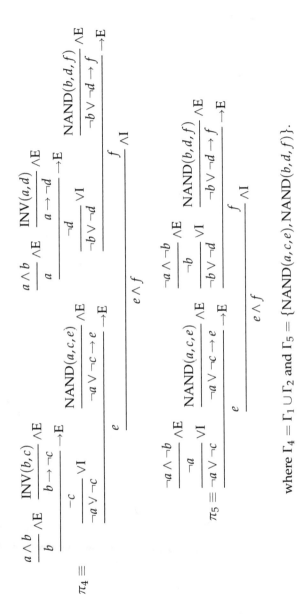

Figure 8.2: The subproofs π_4 and π_5.

- $F_{\pi_3}(\iota, \eta)$ is a function $f : \mathbb{N}^2 \oplus \mathbb{N}^2 \to \mathbb{N}$ such that:

$$f((1,(t_1,t_2))) = \eta^+(F_{\pi_1}((t_1,t_2),\iota,\eta))$$
$$f((2,(t_1,t_2))) = \eta^+(F_{\pi_2}((t_1,t_2),\iota,\eta))$$

- $F_{\pi_1}((t_1,t_2),\iota,\eta) = (1,\eta^-((t_1,\iota^+(t_2)))) \in \mathbb{N} \oplus \mathbb{N}$.

- $F_{\pi_2}((t_1,t_2),\iota,\eta) = (2,\eta^-((t_2,\iota^+(t_1)))) \in \mathbb{N} \oplus \mathbb{N}$.

- $F_{\pi_6}(\iota, \eta)$ is a function $g : \mathbb{N}^2 \oplus \mathbb{N}^2 \to \mathbb{N}$ such that:

$$g((1,(t_1,t_2))) = \eta^-(F_{\pi_4}((t_1,t_2),\iota,\eta))$$
$$g((2,(t_1,t_2))) = \eta^-(F_{\pi_5}((t_1,t_2),\iota,\eta))$$

- $F_{\pi_4}((t_1,t_2),\iota,\eta) = (\eta^+((2,\iota^-(t_2))),\eta^+((2,\iota^-(t_1)))) \in \mathbb{N}^2$.

- $F_{\pi_5}((t_1,t_2),\iota,\eta) = (\eta^+((1,t_1)),\eta^+((1,t_2))) \in \mathbb{N}^2$.

Now, given a concrete stabilisation bound for the INV and the NAND gates we can compute the resulting stabilisation bound for the XOR circuit. Here we consider the stabilisation bounds for INV and NAND given in (8.3) and (8.4); hence $\iota = (f_{INV}, f_{INV})$ and $\eta = (f_{NAND}^-, f_{NAND}^+)$. We get:

$$F_{\Pi}(\iota, \eta) = (F_1, F_2) ;$$
$$F_1((i,(t_1,t_2))) = \begin{cases} \max\{t_1, t_2 + \delta_I\} + 2\delta_N & \text{if } i = 1 \\ \max\{t_2, t_1 + \delta_I\} + 2\delta_N & \text{if } i = 2 \end{cases} ;$$
$$F_2((i,(t_1,t_2))) = \begin{cases} \max\{t_1, t_2\} + \delta_I + 2\delta_N & \text{if } i = 1 \\ \max\{t_1, t_2\} + 2\delta_N & \text{if } i = 2 \end{cases} .$$

As an example, let us suppose that $V(a)$ stabilises to 1 at time 10 and $V(b)$ stabilises to 0 at time 20. The formula $(a \wedge \neg b) \vee (\neg a \wedge b) \to g$ states that $V(g)$ must stabilise to 1, and the stabilisation time is given by the exact stabilisation bound t for V and g. By Theorem 8.3.1, t corresponds to the value of F_1 on the exact stabilisation bound $(1,(10,20))$ for V and $(a \wedge \neg b) \vee (\neg a \wedge b)$; therefore $t = F_1((1,(10,20))) = 20 + \delta_I + 2\delta_N$.

8.4 Concluding remarks

To conclude this section, we show that Theorem 8.3.1 essentially depends on the E calculus and does not hold for proofs of classical logic. Indeed, let us take the formula $\text{XOR}(x,y,z)$ and its disjunctive normal form

$$\text{XOR}'(x,y,z) = (\neg x \wedge \neg y \wedge \neg z) \vee (\neg x \wedge y \wedge z) \vee$$
$$\vee (x \wedge \neg y \wedge z) \vee (x \wedge y \wedge \neg z) \ .$$

Now, $\text{XOR}'(x,y,z)$ is classically equivalent to $\text{XOR}(x,y,z)$ and thus $\text{XOR}'(x,y,z)$ represents the boolean function xor : $\mathbb{N}^2 \to \mathbb{N}$. Moreover, it is easy to find a proof $\Pi' : \mathcal{C}_{\text{XOR}} \vdash \text{XOR}'(a,b,g)$ in the **CL** calculus. On the other hand, as we will show hereafter, there is no $\gamma \in \lceil \text{XOR}'(x,y,z) \rceil$ satisfying Point 1 of Theorem 8.3.1.

First of all, we remark that the set of stabilisation bounds for $\text{XOR}'(x,y,z)$ is isomorphic to $(\mathbb{N}^3 \oplus \mathbb{N}^3 \oplus \mathbb{N}^3 \oplus \mathbb{N}^3)$, thus a stabilisation bound of this set can be written as $(i,(t_1,t_2,t_3))$ with $i \in \{1,\ldots,4\}$ and $t_1,t_2,t_3 \in \mathbb{N}$. Now, let us consider the following stabilisation bounds for the formulae of $\mathcal{C}_{\text{XOR}}$:

- let $\iota = (\iota^-,\iota^+)$ be the stabilisation bound for all the instances of the formula INV(x,y), where $\iota^-(t) = \iota^+(t) = 0$ for every $t \in \mathbb{N}$;

- let $\eta = (\eta^-,\eta^+)$ be the stabilisation bound for all the instances of the formula NAND(x,y,z), with $\eta^-((t_1,t_2)) = 0$ for every $t_1,t_2 \in \mathbb{N}$ and $\eta^+((i,t)) = 0$ for $i = 1,2$ and for every $t \in \mathbb{N}$.

Now, let us assume that $F_{\Pi'}(\iota,\eta) = (1,(c_1,c_2,c_3))$ for some $c_1,c_2,c_3 \in \mathbb{N}$. Let us consider the stable waveform V such that $V(a) = V(b) = V(e) = V(f) = 1$ and $V(c) = V(d) = V(g) = 0$. It is easy to check that

$$\iota, V \models \text{INV}(a,d) \qquad \iota, V \models \text{INV}(b,c)$$
$$\eta, V \models \text{NAND}(a,c,e) \quad \eta, V \models \text{NAND}(b,d,f)$$
$$\eta, V \models \text{NAND}(e,f,g)$$

while

$$(1,(c_1,c_2,c_3)), V \not\models \text{XOR}'(a,b,g)$$

since $(c_1,c_2,c_3), V \not\models \neg a \wedge \neg b \wedge \neg g$.

Similar conclusions are obtained considering $F_{\Pi'}(\iota,\eta) = (j,(c_1,c_2,c_3))$ with $j = 2,3,4$ and $c_1,c_2,c_3 \in \mathbb{N}$.

Bibliography

[1] A. Avellone, M. Ferrari, and P. Miglioli. Synthesis of programs in abstract data types. In P. Flener and K.K. Lau, editors, *Proceedings of the 8th International Workshop on Logic-based Program Synthesis and Transformation*, volume 1559 of *Lecture Notes in Computer Science*, pages 38–45. Springer Verlag, June 1998.

[2] A. Avellone, C. Fiorentini, P. Mantovani, and P. Miglioli. On maximal intermediate predicate constructive logics. *Studia Logica*, 57:373–408, 1996.

[3] H.P. Barendregt. The type free lambda calculus. In *Handbook of Mathematical Logic* [5], pages 1091–1132.

[4] H.P. Barendregt. *The Lambda Calculus: Its Syntax and Semantics*, volume 103 of *Studies in Logic and the Foundations of Mathematics*. Elsevier - North Holland, 2nd edition, 1984.

[5] J. Barwise. *Handbook of Mathematical Logic*, volume 90 of *Studies in Logic and the Foundations of Mathematics*. Elsevier - North Holland, 1977.

[6] M. Benini. *Verification and Analysis of Programs in a Constructive Environment*. PhD thesis, Dipartimento di Scienze dell'Informazione, Università degli Studi di Milano, January 2000.

[7] H. Benl, U. Berger, H. Schwichtenberg, M. Seisenberger, and W. Zuber. Proof theory at work: Program

development in the MINLOG system. In W. Bibel and P.H. Schmitt, editors, *Automated Deduction — A Basis for Applications, Volume II: Systems and Implementation Techniques*, volume 9 of *Applied Logic Series*, pages 41–71. Kluwer Academic Publishers, 1998.

[8] U. Berger and H. Schwicthenberg. Program extraction from classical proofs. In D. Leivant, editor, *Logic and Computational Complexity, International Workshop LCC '94*, volume 960 of *Lecture Notes in Computer Science*, pages 77–97. Springer Verlag, 1995.

[9] C.L. Chang and H.J. Keisler. *Model Theory*. Elsevier - North Holland, 1973.

[10] T.H. Cormen, C.E. Leiserson, R.L. Rivest, and C. Stein. *Introduction to Algorithms*. The MIT Press, 2001.

[11] E.W. Dijkstra. *A Discipline of Programming*. Prentice Hall, 1976.

[12] M. Ferrari. *Strongly Constructive Formal Systems*. PhD thesis, Dipartimento di Scienze dell'Informazione, Università degli Studi di Milano, 1997.

[13] M. Ferrari and C. Fiorentini. A proof-theoretical analysis of semiconstructive intermediate theories. *Studia Logica*, 73:21–49, 2003.

[14] M. Ferrari, C. Fiorentini, and P. Miglioli. Goal oriented information extraction in uniformly constructive calculi. In *Argentinian Workshop on Theoretical Computer Science (WAIT'99)*, pages 51–63. Sociedad Argentina de Informática e Investigación Operativa, 1999.

[15] M. Ferrari, C. Fiorentini, and P. Miglioli. Extracting information from intermediate semiconstructive ha-systems - extended abstract. *Mathematical Structures in Computer Science*, 11(4):589–596, 2001.

[16] M. Ferrari, C. Fiorentini, and M. Ornaghi. Extracting exact time bounds from logical proofs. In A. Pettorossi, editor, *LOPSTR '01: Selected papers from the*

11th International Workshop on Logic Based Program Synthesis and Transformation, volume 2372 of *Lecture Notes in Computer Science*, pages 245–266. Springer Verlag, 2001.

[17] M. Ferrari, P. Miglioli, and M. Ornaghi. On uniformly constructive and semiconstructive formal systems. *Logic Journal of the IGPL*, 11(1):1–49, 2003.

[18] P. Flener, K.K. Lau, and M. Ornaghi. Correct-schema-guided synthesis of steadfast programs. In *Proceedings XIIth IEEE International Automated Software Engineering Conference*, pages 153–160. IEEE, 1997.

[19] P. Flener, K.K. Lau, and M. Ornaghi. On correct program schemas. In N.E. Fuchs, editor, *Proceedings of the 7th International Workshop on Logic-Based Program Synthesis and Transformation*, volume 1463 of *Lecture Notes in Computer Science*. Springer Verlag, 1997.

[20] D.M. Gabbay. *Semantical Investigations in Heyting Intuitionistic Logic*. D. Reidel Publishing Company, Dordrecht, 1981.

[21] J.Y. Girard, Y. Lafont, and P. Taylor. *Proofs and Types*. Cambridge University Press, 1989.

[22] J. Goguen and J. Meseguer. Initiality, induction and computability. In M. Nivat, editor, *Algebraic Methods in Semantics*, pages 459–542. Cambridge University Press, 1985.

[23] J. Goguen and J. Meseguer. Unifying functional, object-oriented and relational programming with logical semantics. In B. Shriver and P. Wegner, editors, *Research Directions in Object-Oriented Programming*, pages 417–477. The MIT Press, 1987.

[24] J.A. Goguen, J.W. Thatcher, and E.G. Wagner. An initial algebra approach to the specification, correctness and implementation of abstract data types. IBM Research Report RC6487, Yorktown Heights, 1976.

[25] C. Kreitz, K.K. Lau, and M. Ornaghi. Formal reasoning about modules, reuse and their correctness. In *Proceedings of the International Conference on Formal and Applied Practical Reasoning*, volume 1085 of *Lecture Notes in Artificial Intelligence*, pages 384–398. Springer Verlag, 1996.

[26] D. MacKenzie. *Mechanizing Proofs*. The MIT Press, 2004.

[27] M. Mendler. A timing refinement of intuitionistic proofs and its application to the timing analysis of combinational circuits. In P. Miglioli, U. Moscato, D. Mundici, and M. Ornaghi, editors, *Proceedings of the 5th Workshop on Theorem Proving with Analytic Tableaux and Related Methods*, volume 1071 of *Lecture Notes in Artificial Intelligence*, pages 261–277. Springer Verlag, 1996.

[28] M. Mendler. Timing analysis of combinational circuits in intuitionistic propositional logic. *Formal Methods in System Design*, 17(1):5–37, 200.

[29] M. Mendler. Characterising combinational timing analyses in intuitionistic modal logic. *Logic Journal of the IGPL*, 8(6):821–852, 2000.

[30] P. Miglioli, U. Moscato, and M. Ornaghi. PAP: a logic programming system based on a constructive logic. In M. Boscarol, L. Carlucci Aiello, and G. Levi, editors, *Foundations of Logic and Functional Programming*, pages 143–156. Springer Verlag, 1988.

[31] P. Miglioli, U. Moscato, and M. Ornaghi. Semi-constructive formal systems and axiomatization of abstract data types. In J. Diaz and F. Orejes, editors, *TAPSOFT'89*, volume 351 of *Lecture Notes in Computer Science*, pages 337–351. Springer Verlag, 1989.

[32] P. Miglioli, U. Moscato, and M. Ornaghi. Program specification and synthesis in constructive formal systems. In K.K. Lau and T.P. Clement, editors, *Logic Program Synthesis and Transformation, Manchester 1991*, pages 13–26. Springer Verlag, 1991.

[33] P. Miglioli, U. Moscato, and M. Ornaghi. Abstract parametric classes and abstract data types defined by classical and constructive logical methods. *Journal of Symbolic Computation*, 18:41–81, 1994.

[34] P. Miglioli, U. Moscato, and M. Ornaghi. How to avoid duplications in refutation systems for intuitionistic logic and Kuroda logic. In K. Broda, M. D'Agostino, R. Goré, R. Johnson, and S. Reeves, editors, *Theorem Proving with Analytic Tableaux and Related Methods: 3rd International Workshop, Abingdon, U.K.*, pages 169–187, May 1994.

[35] P. Miglioli, U. Moscato, and M. Ornaghi. Avoiding duplications in tableau systems for intuitionistic and Kuroda logics. *Logic Journal of the IGPL*, 1(5):145–167, 1997.

[36] P. Miglioli, U. Moscato, M. Ornaghi, and G. Usberti. Constructive validity and classical truth to assign meaning to programs. In *Proceedings of the Second World Conference on Mathematics at the Service of Man*, pages 490–500. Universidad Politecnica de Las Palmas, 1982.

[37] P. Miglioli, U. Moscato, M. Ornaghi, and G. Usberti. Constructive validity and classical truth to assign meaning to programs. Technical Report MIG-10, Istituto di Cibernetica, Università degli Studi di Milano, 1982.

[38] P. Miglioli, U. Moscato, M. Ornaghi, and G. Usberti. A constructivism based on classical truth. *Notre Dame Journal of Formal Logic*, 30(1):67–90, 1989.

[39] P. Miglioli and M. Ornaghi. A purely logical computing model: the open proofs as programs. Technical Report MIG-7, Istituto di Cibernetica, Università degli Studi di Milano, 1978.

[40] P. Miglioli and M. Ornaghi. A logically justified model of computation I. *Fundamenta Informaticæ*, IV(1):151–172, 1981.

[41] P. Miglioli and M. Ornaghi. A logically justified model of computation II. *Fundamenta Informaticæ*, IV(2):277–341, 1981.

[42] P. Miglioli and M. Ornaghi. Constructive proofs and logical computations. *Pocitace umela inteligencia*, 1(5):369–388, 1982.

[43] Motorola Inc., editor. *MC68020 32-bit Microprocessor User's Manual*. Prentice Hall, New Jersey, 1989.

[44] D. Nelson. Recursive functions and intuitionistic number theory. *Transactions of the American Mathematical Society*, 61:307–368, 1947.

[45] P. Odifreddi. *Classical Recursion Theory*, volume 125 of *Studies in Logic and the Foundations of Mathematics*. Elsevier - North Holland, 1989.

[46] H. Schwichtenberg. Proofs as programs. In P. Aczel, H. Simmons, and S.S. Wainer, editors, *Proof Theory. A selection of papers from the Leeds Proof Theory Programme 1990*, pages 81–113. Cambridge University Press, 1993.

[47] H. Schwichtenberg and A.S. Troelstra. *Basic Proof Theory*, volume 43 of *Cambridge Tracts in Theoretical Computer Science*. Cambridge University Press, 1996.

[48] A.S. Troelstra. *Metamathematical Investigation of Intuitionistic Arithmetic and Analysis*, volume 344 of *Lecture Notes in Mathematics*. Springer Verlag, 1973.

[49] A.S. Troelstra. Aspects of constructive mathematics. In *Handbook of Mathematical Logic* [5].

[50] A.S. Troelstra and D. van Dalen. *Constructivism in Mathematics: An Introduction I and II*, volume 121, 122 of *Studies in Logic and the Foundations of Mathematics*. Elsevier - North Holland, 1988.

[51] A.S. Troelstra and D. van Dalen. *Constructivism in Mathematics: An introduction. Volume 1*, volume 121 of *Studies in Logic and the Foundations of Mathematics*. Elsevier - North Holland, 1988.

Index

www.ingramcontent.com/pod-product-compliance
Lightning Source LLC
LaVergne TN
LVHW011940060326
832903LV00045B/34